What Use Is Chemistry?
An A to Z History of Amazing Chemical Technologies

John P.H.M. Blackie

ISBN: 978-1-83563-468-4

www.newgeneration-publishing.com

New Generation Publishing

'The most important of my discoveries have been suggested to me by my failures'

Sir Humphry Davy FRS (1778–1829) was a pioneering British chemist and inventor who discovered many new elements and became a great exponent of the scientific method and a pioneer of electrochemistry.

Foreword

By an experienced business leader and expert in the biotechnology and chemical industries.

My mother was a scientist whose passion was its application in the home. She lectured in Science for Home Economics and was an early contributor to the care labels we all take for granted on our clothes. Washing powders, refrigeration and the benefits of insulating a house, to name but a few, were in her eyes all the obvious benefits of science to be welcomed unashamedly as they made our lives easier and more affordable.

Growing up in this world I took to science at school and in particularly chemistry – I was intrigued by the way in which you could predictably change substances through reactions with other substances and that by careful design of a series of reactions you could create complex molecules that had characteristics and properties that made them useful in our lives.

A chance encounter with chemical engineers at a careers evening led me to discover the world of 'scale-up' and as my chemistry teacher was fond of telling us the 'sordid world of using pressure and temperature' in other words, the chemical industry.

Forty years and more later I have had a career that as I turned the pages of John's book and read his A to Z, I am proud to have played a little part in using some of the amazing discoveries and developments he describes. As his colleague we made dyestuffs, pigments, and agrochemicals together – then as we went our separate ways, I entered the world of biochemical manufacturing – making pharmaceuticals, advanced therapies and vaccines in ways that could not have been done when I started my career. Initially I qualified as a chemical engineer out of Manchester, but my journey took me around the world through manufacturing management, materials procurement, supply chain management, business development, sales, marketing and eventually to the joys of general management and the lows and highs of running a contract manufacturing and development company. Nowadays I sit on Boards and work with companies to innovate and use the technologies John describes so well in this book.

As he looks to the future in each chapter – it is clear, that our priorities as a society are continuing to change. What was seen as an economic benefit and necessity 50 or even 20 years ago is now balanced with a real understanding of the impact we are having on nature, biodiversity and climate. Scientists and engineers have a huge role to play in allowing us to continue to enjoy the benefits of the technical developments we have made whilst ensuring a liveable future. Whatever age you are reading this book – there is so much more to do – please use John's accessible stories to either inspire yourself or others to be curious and find out more.

Steve Bagshaw CBE

Contents

1. Introduction

Many may ask *'What use is chemistry?'* Chemical technology often goes unnoticed in the background of people's daily lives, but its application has contributed significantly to global advancement, with the constant rearrangement of atoms, elements and molecules and exchanges of energy that are at the core of the chemical sciences. The chemical, pharmaceutical and a wide range of other industries have employed chemical technologies to carry out useful chemical reactions and processing, converting raw materials into new products that have truly changed the world.

The history of advances in chemical technology can first be seen in efforts made by primitive metallurgists that ended the Stone Age, bringing an age of bronze and iron. Those civilizations that later mastered iron making had a huge advantage in warfare and material progress. In the Middle East, artisans refined alkalis made from ash and limestone to produce glass in the 7th century BC, the Phoenicians dyed fabrics with Tyrian purple from snails and made soap in the 6th century BC, while the fermentation of fruit and milk played an important role in Ancient Egypt. Chemical technology in the past often facilitated progress in other aspects of life. Salt was used to preserve and flavour meats and fish. About 100 BC, the papermaking process was developed in China – which allowed its vast empire to be governed using written communications. Later in the Islamic world, scholars were able to record and communicate knowledge from past civilizations. Chinese alchemists during the 10th century made gunpowder for fireworks and improved weaponry which spread to the West, and elsewhere others attempted to transform base metals into gold. The early chemists who emerged in the Middle Ages created primitive medicines, pigments and dyes.

Recreation of a 16th century alchemists laboratory in Basel

One of Britain's greatest contributions to the modern world was its initiation of the Industrial Revolution from the mid-18th century. Britain greatly benefited when its entrepreneurs applied new sciences and engineering, including chemical technology, to exploit its ready availability of raw materials, such as coal, iron ore and salt, combined with access to financial capital, stable government and free-market trade. From industrialization came new sources of energy, agricultural improvements, increasingly sophisticated metals and multiple inventions that transformed productivity, transportation and living conditions.

In 1840 coal-fired steam engines in Britain were generating 40% of the motive power in use across Europe. Britain retained its industrial leadership into the late 19th century when Germany and the United States (US) were able to build on Britain's pioneering advancements. The British population had expanded explosively, from 6 million in 1750 to reach 42 million by the turn of the 20th century. This growth meant many more workers moved to towns, needing food, homes, consumer goods, clean water and sanitary facilities. Rural life changed too because factory-made tools and machinery combined with the selective breeding of crops and animals increased farm productivity. From the 1830s the application of chemical fertilizers began to improve crop yields.

Over time, advances in chemical manufacturing allowed the bulk production of new acids and alkalis, steelmaking, improved cements, synthetic dyestuffs, powerful explosives, water treatment and enhancements in health from synthetic medicines. These advances saw Britain become a great trading nation with a worldwide empire covering a quarter of the globe.

Innovations in chemical technology spread across the developed world from the early 20th century driven by the availability of petroleum-based fuels, the distribution of electricity and nitrogen fertilizers and improvements in public sanitation. New chemical technologies supported the development of antibiotics, preventative vaccines and other effective medicines, plastics for consumer goods, cars and airplanes, radio and television, spacecraft and lasers, medical imaging and electronic computers. The latter decades of the 20th century saw new innovations in silicon chemistry driving the rise of semiconductors, bringing digital and mobile technology. Chemical technologies have been instrumental in creating new high-performance materials, medicines based on genetic science and biotechnology, processed foods, environmental abatement, recycling of waste and renewable sources of energy.

Although the chemical industry proved critical for Britain's economic success and enabled improvements in virtually all sectors of the economy, and Britain pioneered many new inventions, it is disappointing that chemical technologies are often neglected in history books. There is a focus on mechanical innovation such as steam engines, textile machinery, civil feats of engineering, bridges and the railways. Then discoveries in physics, electricity, aviation, nuclear energy, space rockets and of course electronics and computing are also celebrated.

The chemical technologies at the heart of oil processing and the making of polymers, nitrogen fertilizers, batteries, a rainbow of colourants or antibiotics may get only a passing mention. There are overlaps as chemistry is critical to steelmaking and silicon chip manufacture. It was synthetic organic chemistry that led to dyes, then medicines and plastics. The early 20th century saw the revolutionary process to fix nitrogen from air and the invention of that essential plastic polyethylene. Chemistry played an important role in World War II, but the contribution of science is normally discussed in terms of the physicists and engineers who produced atomic bombs, advanced aircraft or engines, radar and rockets. In fact, the chemical industry was vital to the war effort in creating processes for hydrogenating coal into gasoline, refining oil into aviation fuel and making polymers for electrical insulation and aircraft parts, while providing anti-malaria medicines and antibiotics to prevent the deaths of wounded troops. The industry also made possible synthetic rubber for tyres and engine seals. Post-war all these advances proved to be useful.

Nowadays, with pressure on delivering set curriculums, the histories of most chemical technologies have been written out of secondary school courses and text books. Hopefully this book showcases some of the most important traditional chemical (with a few biochemical) technologies from the past few centuries, profiling the pioneers who evolved them, their achievements and the ways these technologies have influenced the modern-day world. Inclusion in the book is a result of a broad analysis of their impacts, but the writer has also tried to demonstrate the fantastically wide variety of chemical technologies we rely on. While arranged in a logical A to Z sequence, they can of course be read in any order or dipped in and out of.

It is modestly hoped this book will inspire chemical technologists of the future while helping you see how chemical technology has been influential in the history of industrial, social and economic development. After all, when we become interested in what we see around us – when we can start to see how we might become involved in developing a new chemical technology someday – then the seeds of a future passion are sown.

Note: *All commercial information provided in the book is based on approximate market sizes, turnovers and product values (in US$ or GB£) taken from publicly available sources within the period of 2020 to 2023 unless otherwise stated.*

2. The Chemical Industry

The modern chemical industry is a vast collection of different activities. It is broadly described as the industry that uses chemical technology to manufacture chemical products but includes any reliance on chemical sciences. Over the past 250 years the chemical industry has become vital in the modern global economy. Some 95% of all manufactured goods are dependent in some way on the chemical industry. Globally the industry has the capability to produce hundreds of thousands of chemicals for end consumers or intermediaries. The products of the chemical industry nearly always undergo further processing before reaching their ultimate consumer. About 25% of its output is consumed within the industry itself. A typical chemical product is passed from facility to facility several times before it emerges into the market to be used in another industry. Commodity chemicals, such as methanol, chlorine, alkalis or sulfuric acid, are used in numerous intermediates. There can be multiple uses for a particular chemical. For example, the largest use for ethylene glycol is as an antifreeze, but it is also used as a hydraulic brake fluid. Its further processing leads to PET (polyethylene terephthalate) polymers for plastic bottles, additives for the textile, pharmaceutical and cosmetic industries or emulsifiers for formulating insecticides and fungicides.

The chemical industry has solved pressing problems, by synthesizing agricultural fertilizers and chemical pesticides to ensure there is a growing food supply, eradicating deadly diseases by producing life-saving pharmaceuticals, brightening the world with synthetic dyes and pigments, producing high-performance fibres for sports and fashion, and making coatings and paints to protect built structures. It has made possible the synthetic sweeteners, flavours and additives that enhance foods, soaps for hygiene and beauty products for the cosmetics industry, innovative materials for electronics, cars and packaging and artificial rubber for tyres, not to mention a vast range of other synthetic materials used in industrial, construction and consumer products.

Pilot-scale fine chemicals plant, Grangemouth

The global chemical industry produces over £4,000 billion-worth of core chemical products, not including those heavy industries concerned with chemistry applied to steel and concrete manufacturing. There are millions employed worldwide in the chemical, pharmaceutical and associated industries. They work in research and development, design, operations, testing or supporting, or selling to customers. Many are qualified chemists, biochemists, chemical engineers, technicians, laboratory analysts and functional specialists. It is a multinational industry, situated in multiple locations. The largest British based chemicals group is INEOS while German company Badische Aniline und Soda Fabrik (BASF) is the largest globally, having major sites across Europe, the US, China, Southeast Asia and at hundreds of other locations worldwide. Overall, China is now the world's leading supplier of chemicals, minerals and metals. It is forecast that global chemical sales will grow to over £5,500 billion by 2030.

All companies constantly face competition and must innovate to remain successful. New ways are being found for the chemical industry to satisfy its increasingly sophisticated, demanding and environmentally conscious consumers. Many chemicals are potentially hazardous, whether flammable, explosive, corrosive or toxic. Manufacturing processes frequently involve high temperatures, high pressures and reactions that can be dangerous unless carefully controlled. In the past, chemical production has been occasionally accompanied by environmental pollution, natural resource depletion and adverse health impacts, but nowadays the industry operates within stricter safety and environmental limits demanded by national legislation. The most important considerations when designing modern chemical plants are the safety of processes and products, energy efficiency, mitigating climate change and effective waste management. Companies spend large amounts of money on safety programmes to ensure that chemicals do not adversely affect human health or the environment. Modern plants are designed to minimize pollution and protect people and biodiversity by reducing emissions and groundwater or soil contamination. The chemical industry has expanded its role into other industries to achieve resource and energy efficiency by providing technologies that support decarbonization, energy saving, material recycling, land reclamation and sustainable development. The industry provides many high-value jobs, technically interesting and exciting opportunities to contribute to changing the world for the better.

3. Future Challenges

It is easy to argue that the world needs more chemists right now to help ensure survival and success in the future. Chemistry is playing a key role in tackling today's challenges and ensuring a sustainable future. During the 2020 pandemic, biochemistry was at the fore with messenger RNA, lipid nanoparticles and antivirals, for example. Adapting to the impact of climate change on the environment and reaching net zero carbon emissions require a significant reduction in the use of fossil fuels, particularly for transport, heating and industrial processes. There are many technological challenges to be addressed during the transition – and a great many of these require the knowledge and skills of chemists and chemical engineers.

Globally each year, chemical and petrochemical industries extract and transform over 1,300 million tonnes of raw materials derived from fossil fuels and use some 600 other raw materials and 300 million tonnes of water. Only about a third of those inputs are transformed into useful products as the vast majority are lost as waste or emitted into the environment. Only 20% of what is produced is recovered and reused, so chemical technologists must discover ways to improve processes to reduce the impact on the environment and our own health and to create a more 'circular' economy. Some solutions may come from emerging technologies, such as functionalized zeolites and metal–organic frameworks for carbon capture. Or maybe carbon dioxide will be employed as an important raw material. The chemical sector is the largest industrial energy consumer and a large direct carbon dioxide emitter. However, it must be appreciated that half of the input is consumed as feedstock – fossil fuels used as a raw material rather than as a source of energy. There is growing demand for a vast array of chemical products, including plastics and primary chemicals.

Some hydrocarbon fuels are likely to continue to be required for aviation jet fuel in the medium term due to the excessive weight penalties associated with current alternatives such as batteries. New biofuels produced from biomass or synthetic fuels derived from atmospheric carbon dioxide and 'green' hydrogen are attractive alternatives. Green hydrogen is not produced from converting natural gas but is derived from the electrolysis of water. Hydrogen is also essential for the synthesis of ammonia – the feedstock for nitrogen-based fertilizers – but the process uses up to 2% of global energy production and generates considerable carbon dioxide. The sustainable production of ammonia will have to replace the current process to fix nitrogen. Can chemical technologists of the future find a new way to keep increasing food production as the global population continues to grow?

Donaldsonville ammonia plant in Louisiana

An electrically powered world needs major improvements in the lifespan of batteries and fuel cell reliability. Chemists have previously helped develop many exciting discoveries in materials science, such as new alloys, low-energy LED-based lighting technologies, energy storage using lithium-ion batteries and components for renewable energy, such as perovskite-based photovoltaic (PV) cells, which now cost 250 times less and produce over a third of all electricity. These and other important advances have allowed a more than doubling of the economic output per unit of energy used since the 1990s. But some major challenges remain in the renewable energy field. More batteries will be required in ever larger quantities for electric vehicles, but these currently require lithium and cobalt, elements where known reserves are finite or often obtained from conflict zones. There is a need to develop new batteries using more earth-abundant elements. Similarly, it is hard to imagine life without the benefits of smartphones and electronic computing, but their manufacture requires the use of many rare elements – from sources that are expected to be exhausted in less than 100 years. Chemical technologists need to develop large-scale recycling of existing electronic materials along with creating new materials that employ more abundant elements.

Plastics and composites are now ubiquitous and find application in just about every aspect of modern society. Sometimes these are difficult to recycle or are too dependent on oil. In the future, biodegradable plastics and recyclable composite materials with extended lifetimes are needed to reduce waste. For some carbon-using chemical processes it will be difficult to phase out the use of fossil fuel resources.

The boundary between chemistry and biology will see new technologies with promising applications in medicine and pharmacology. Producing new life-saving treatments for emerging and genetic diseases, drug-resistant superbugs or lifestyle ailments are priorities. Technologies based on mechanochemistry, microbiome studies and targeted protein degradation are possible. The first enables new transformations without solvents, making chemistry greener and giving access to novel reactivities and selectivity. The analysis of the gut microbiome could lead to better understanding of many chronic diseases, as well as reveal interesting compounds for new treatments. Finally, understanding protein chemistry may lead to innovative therapies for diseases such as multiple sclerosis, Parkinson's or Alzheimer's.

Chemists are adopting digital tools such as artificial intelligence (AI), robotics, 'big data' and supercomputing – which will lead to breakthroughs and free chemists from more mundane or repetitive tasks, leaving more time for creative work and collaborations with other disciplines. There are opportunities to find smaller-scale chemical manufacturing methods to transform low-income countries by giving them access to essential chemicals, clean water, pharmaceuticals and sustainable energy solutions.

To achieve all this and more will need more chemists, chemical engineers and others skilled in chemical or biochemical technology to address the global challenges. Chemistry will have a central role as the basis for the necessary research, development and creating practical solutions.

The history of chemical technology is one of solving problems, creating new opportunities and making life better. This is widely illustrated in the pages that follow. There has never been a more exciting and important time to pursue a career in chemistry to contribute to addressing the big challenges that will most certainly impact everyone's lives for the next 100 years.

4. An A to Z of Chemical Technologies

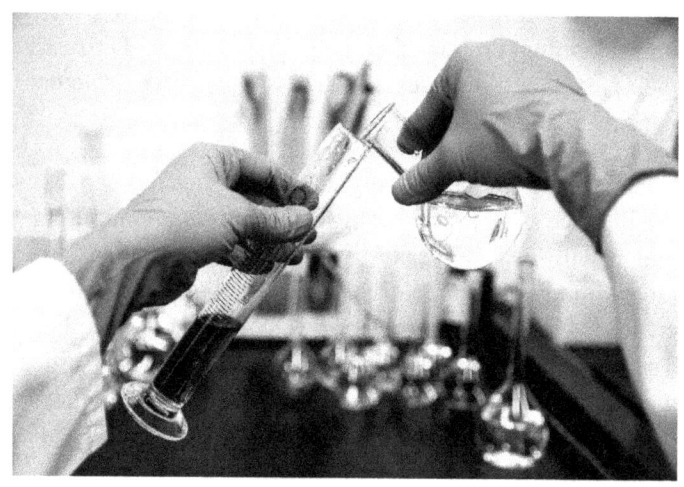

A: Acids and alkalis

Tennant's St. Rollox Chemical Works, Glasgow (1831)

Birth of the chemical industry

The chemical technologies for making inorganic acids and alkalis mark the very beginnings of the modern chemical industry. They grew alongside the many new developments that proved vital to the Industrial Revolution. Sulfuric acid was used in pickling iron (removing rust) in metalworking and for treating textiles. From the mid-18th century, it was made from the new *lead chamber* process. Before that it had been obtained from roasting and distilling naturally occurring copperas or green vitriol – the ancient name for hydrated ferrous or iron sulfate – in a retort. The second important industrial chemical that came to the fore was the alkali sodium carbonate or soda ash, required for manufacturing soap and glass and bleaching textiles or paper. From the early 19th century, soda ash production was transformed by the development of the *Leblanc soda* process.

History

The Romans had referred to sour substances, such as vinegar or lemon juice, as acids (Latin: *acidus* meaning sour). The early alchemists used strong acids to corrode metals and dissolve certain minerals. To produce alkalis various ashes were used. In the Middle East an alkali (Arabic: *al-qali)* meant material obtained from the ashes of the glasswort plant. The Greeks had taken ashes from burning barilla plants, seaweed or kelp, then mixed them with animal fat to make a primitive soap. The water-extract of ashes, called *potash (pot ash)*, composed mostly of potassium carbonate had been used since the Bronze Age in bleaching textiles and making glass, ceramics and soap. Later potash became essential for producing fertilizers and was a traded commodity. After heating potash with slaked lime (calcium hydroxide), a far more alkaline substance, known as caustic potash (potassium hydroxide), was obtained.

Around 1300 a Spanish alchemist, Arnaldus de Villa Nova, used an extract from a lichen, called litmus, for studying acids and bases. The English chemist Robert Boyle (1627–1691) found that certain botanically derived substances, such as syrup of violets, changed colour in the presence of acids or alkalis. These methods allowed chemists to work out which proportion of acids and bases of different strengths would neutralize each other. Chemists began to define bases as substances that could neutralize acids to form water and a salt.

It was Antoine Lavoisier (1743–1794), the brilliant French chemist, who first attempted a more systematic study of acids and bases. He proposed that the presence of oxygen was responsible for an acid's unique properties. The British chemist Humphry Davy (1778–1829) discovered that oxygen was not present in some acids and so could not be the sole element responsible. The German chemist Justus von Liebig (1803–1873) reasoned that hydrogen was the element common to all acids. Later, Swedish chemist Svante Arrhenius (1859–1927) proposed that they gained their properties because of the action of ions in the solutions by adding hydrogen cations [H^+]. From the 18th century, attention turned to how to make acids and alkalis commercially on a larger scale.

Lead chamber process

The German chemist Johann Glauber (1604–1670) had prepared sulfuric acid by burning natural sulfur (brimstone) together with saltpetre (potassium nitrate) in the presence of steam. As the saltpetre decomposed, it oxidized the sulfur into sulfur trioxide, which when combined with water produced sulfuric acid. Joshua Ward (1685–1761), a London pharmacist, began the small-scale production of sulfuric acid using glass vessels at a factory in Twickenham, but its unpleasant smell greatly displeased nearby residents. His factory came to be known as the 'Great Vitriol Works' and was forced to close. The first successful commercial-scale attempt to produce sulfuric acid was made in 1746 by John Roebuck (1718–1794), using his *lead chamber* process. Roebuck was born in Sheffield and had studied medicine at Edinburgh, where he developed an interest in chemistry from attending the lectures of William Cullen and Joseph Black. Later he graduated from the University of Leiden, then started a medical practice in Birmingham while devoting much of his spare time to chemistry.

Roebuck greatly increased the scale of the manufacture by replacing the expensive and fragile glass vessels with larger, less expensive chambers made of riveted sheets of lead. His process involved igniting sulfur and potassium nitrate in a room lined with lead. The saltpetre oxidized the sulfur, which, as the floor of the room was filled with water, reacted to produce sulfuric acid. In 1749 he built what can be considered the first successful chemical factory in Britain at Prestonpans, in east Scotland. His sulfuric acid process enjoyed a monopoly for some years before others copied his method. When the price of raw sulfur increased, the mineral iron pyrite (iron sulfide) was roasted in air to generate the sulfur dioxide, while the iron oxide by-product was consumed in his iron works.

In 1835 Joseph Gay-Lussac (1778–1850) improved the original process by recovering the noxious nitrogen monoxide and recycling it to replace the saltpetre as an oxidizing source. This reduced emissions and cut dependence on scarce saltpetre. The saltpetre was originally sourced from nitre beds fed with manure but became more widely available with the discovery of Chilean saltpetre. A system circulating water in counter-current absorption towers (Glover and Gay-Lussac towers) allowed both heat and material exchanges to take place continuously. It was the first continuous chemical process and the first to employ a homogeneous (same phase) catalyst. The crude sulfuric acid was a dirty brown oil of vitriol (BOV) of 70–78% strength, which after distillation was made into rectified oil of vitriol (ROV) of 96% strength.

The lead chamber process was eventually displaced by the far more economical *contact process* patented by Peregrine Phillips, a British vinegar merchant, in 1831. This process employed a heterogeneous (different phase) catalyst – originally platinum. The process begins by forming sulfur dioxide, in the presence of the catalyst. Sulfur dioxide and air then react enthalpically to give sulfur trioxide [$2 SO_{2(g)} + O_{2(g)} \rightleftharpoons 2 SO_{3(g)}$]. By Le Chatelier's principle, a lower temperature should be used to shift the chemical equilibrium to increase the product yield, but a lower temperature also lowers the rate of reaction to an uneconomical level. Instead, to increase the rate, a higher temperature (450 °C) with medium pressure (1 to 2 atmospheres) are used to ensure an adequate conversion yield. The hot sulfur trioxide formed is added to sulfuric acid in an absorption tower, which generates highly concentrated oleum acid [$H_2S_2O_7$], which is in turn added to water to form usable sulfuric acid. Purification of the air and sulfur dioxide streams is necessary to avoid catalyst poisoning. As the original catalyst was susceptible to poisoning from arsenic impurities in sulfur it was replaced by one of vanadium pentoxide. The sulfuric acid process was of paramount importance in the early British chemical industry because the acid had multiple uses – being strong and corrosive, an oxidizing agent and a dehydrating agent. Because of its application in the production of many manufactured chemicals and goods, Justus von Liebig famously wrote in 1843: *'It is no exaggeration to say that we may judge the commercial prosperity of a country from the amount of sulfuric acid that it consumes.'*

Sodium carbonate

After sulfuric acid the next essential inorganic chemical to be produced on an industrial scale was sodium carbonate, commonly known as soda ash. It also had multiple uses in many industries, including to manufacture calcium hypochlorite or calcium oxychloride for bleaching textiles and paper. Historically, the cleansing and bleaching of cloth was achieved by the process of soaking it in urine or exposing the cloth to sun in 'bleachfields'. As the population increased, demand for alkali grew, which further drove demand for textiles, dyeing, soap and glass. Potash (potassium carbonate) alkali was traditionally sourced from hardwood or other plant ashes. In the 1720s, alkali was being produced in the Scottish Orkney islands by burning seaweed. When dried it formed brittle purple kelp; it required 20 tonnes of seaweed to produce just one tonne of kelp. This employed up to 3,000 islanders throughout the summertime.

An early chemical process for making soda ash was established in 1780 by Scottish chemist James Keir (1735–1820) in conjunction with Alexander Blair at Tipton in the West Midlands. The method required seeping a weak solution of sulfuric acid through a thick sludge of lime. The acid reacted with the lime, forming sodium sulfate, leaving the alkali, calcium sulfate, to run off as a clear liquid. The Tipton works eventually converted waste sodium and potassium sulfates into alkali for soap boiling. As demand for soda ash grew, the French chemist Nicolas Leblanc (1742–1806) made the most important advance in the chemical technology for this alkali, which laid the foundations for many industrial-scale chemical plants in the future.

Pioneer

Nicolas Leblanc was born in 1742 near Orléans in France. Orphaned by the age of 9, he was raised by a local doctor and sent to study surgery in Paris. This brought him into the service of Louis Philippe, the Duke of Orléans, who provided him with an income and spare time to study chemistry. France experienced a shortage of alkali as the supply of both barilla ashes and wood ash were impacted by the demand from shipbuilding for timber because of war. In 1783 King Louis XVI and the French Academy of Science offered a reward for the development of an economical process to manufacture soda from sea salt rather than ashes. Leblanc began his research by mixing sodium chloride with sulfuric acid in a cast-iron pan heated to 800 °C to produce sodium sulfate and hydrogen chloride gas [$2NaCl + H_2SO_4 \rightarrow 2HCl + Na_2SO_4$]. He found that adding chalk or limestone (calcium carbonate) to some of the crushed sodium sulfate mixed with charcoal then heating it in a furnace to 1,000 °C gave a molten mixture of sodium carbonate and calcium sulfide [$Na_2SO_4 + CaCO_3 + 2C \rightarrow Na_2CO_3 + CaS + 2CO_2$]. Water was added to the mixture (black ash) to separate the soluble sodium carbonate from the insoluble waste calcium sulfide, after which it was treated with carbon dioxide to remove impurities and finally the carbonate solution was evaporated to dryness in open pans leaving pure soda crystals behind. While Leblanc did not fully understand the complex reactions taking place inside the furnace, the result was an economically viable industrial production process for pure soda from the easily obtainable raw materials of sea salt, sulfuric acid, limestone and charcoal.

He scaled up the process, obtained a patent and then built the first full-scale soda plant at Saint Denis, near Paris. Unfortunately, he was to suffer several serious setbacks.

The chaotic French Revolution in 1789 cut his supplies of saltpetre for sulfuric acid because it was commandeered by the revolutionary government to produce gunpowder for its forces. In late 1793, Leblanc's sponsor, the Duke of Orléans, was accused of being a royalist sympathizer and sent to the guillotine. The revolutionary government seized control of Leblanc's soda factory; it ordered him to reveal the details of his process and evicted him from the factory. Without any income Leblanc struggled to support his family, but he continued to campaign for the return of the factory and restoration of his exclusive rights to the soda ash process. Finally, in 1804 he was awarded some compensation for the assets that had been seized but this was nowhere near sufficient for him to restart his factory. Tragically, Leblanc, a broken and depressed man, ended his own life in 1806.

Leblanc's soda ash process did survive the revolution, reaching an annual output of 10,000 tonnes by 1818, but it was in the United Kingdom (UK), with a more enlightened government that encouraged entrepreneurs, that the process was able to prosper on a large scale. The first UK soda works using the Leblanc process was built by the chemist William Losh (1770–1861) on the River Tyne in 1816. The plants were environmentally noxious, producing large amounts of hydrochloric fumes from chimneys and smelly solid waste (calcium sulfide) from the black ash extraction stage, although at that time pollution would have been much less of a concern. About 5.5 tonnes of acid gas and 1.75 tonnes of solid waste were generated for every tonne of soda made. The main alkali soda works were sited in Tyneside, Cheshire, Merseyside and Glasgow, all of which had easy access to the essential raw materials of salt, coal and limestone. Many strands of the early Industrial Revolution came together where key waterways or canals converged. A town such as Widnes in Cheshire thrived as a centre for the early chemical industry. It had neighbouring coal fields, proximity to the rock salt deposits, with soap and glassmaking industries at St Helens. An illustrious number of pioneering industrial chemists had factories sited there to produce soda ash, but also soap, borax, salt cake and bleaching powder. These new chemical technologies laid the foundation for much of Britain's early chemical industry. James Muspratt (1773–1886) in Merseyside and Charles Tennant (1768–1838) in Glasgow soon established some of the largest chemical plants in the world. By 1840 synthetic soda had largely replaced natural soda from ash. British soda production dwarfed that of France, eventually reaching more than 200,000 tonnes annually before the Le Blanc process itself was made obsolete by the more environmentally friendly *Solvay* process.

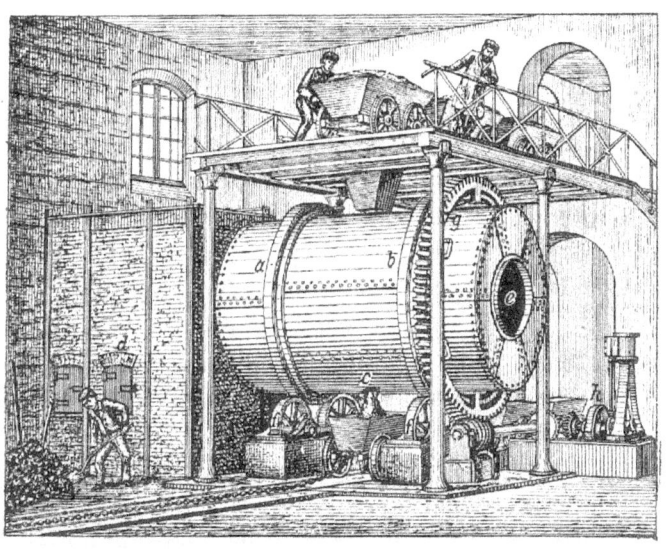

Leblanc soda cylinder furnace in the 19th century

Many areas became heavily polluted; Widnes town was once described as *'the dirtiest, ugliest and most depressing town in England'*. The acid gas killed trees, damaged buildings and caused nuisance to its neighbours, while the solid waste blighted the landscape for miles around. The government was forced to pass one of the earliest pieces of air-pollution regulation (The Alkali Act of 1863) that required a 95% reduction in hydrogen chloride emissions. This was achieved at first by passing the emissions through coke-packed water 'scrubbing' *Gossage* towers, converting the hydrogen chloride to hydrochloric acid. Unfortunately, this resulted in water pollution when towers were discharged into local waterways. The problem was solved more effectively in 1874, when Henry Deacon in Widnes developed the process for catalytic oxidation of hydrogen chloride to chlorine, a product much in demand for manufacture of bleaching powder in the paper and textiles industries. The unpleasant solid black ash waste of calcium sulfide was deposited in landfill sites near to the factories, but in time it reacted with water to produce sulfur dioxide and hydrogen sulfide, giving off a stench of rotten eggs and posing a significant health hazard to residents. The waste was not only environmentally offensive; it resulted in a loss of valuable sulfur from the lead chamber acid process. From 1887 it was possible to recover the sulfur with the introduction of the *Chance and Claus* process. The treatment of the waste with carbon dioxide and combustion of the resulting hydrogen sulfide in a limited supply of air allowed recovered sulfur to be recycled into the Leblanc works or sold on to other sulfuric acid manufacturers. A new market for pure sulfur had emerged because of the establishment of the *contact process* for sulfuric acid, which grew with creation of the synthetic dyestuffs industry (see D).

Solvay process

It was the Belgian chemist and industrialist Ernest Solvay (1838–1922) who finally solved the significant problems of the Le Blanc soda works in 1861. His ammonia-soda process involved the reaction of carbon dioxide with an ammonia-saturated solution of salt to produce soda ash (sodium carbonate) from brine (sodium chloride) and limestone (calcium carbonate). The ammonia was used in one reaction but reclaimed in a later step so that only a relatively small amount of ammonia was consumed. When properly designed and operated, Solvay's plant producing sodium carbonate reclaimed almost all the ammonia, meaning the only major inputs to the process were brine, limestone and heat, while the only major by-product was calcium chloride, which was sold as road salt.

By 1865 Solvay had established a large-scale plant in Belgium overcoming problems of gas handling and absorption. The distinguished German industrial chemist Ludwig Mond collaborating with Solvay and in partnership with British industrialist John Brunner (1842–1919) began making alkali at their factory at Winnington in Cheshire. The site was located near a river, having access to supplies of Northwich salt and Buxton limestone. The ammonia-soda process prospered as it was more economic and cleaner, which allowed production of soda ash to increase rapidly.

In the UK, as the original Leblanc producers found the new Solvay competition intense, they amalgamated to form the United Alkali Company (UAC) in 1891. Their relief was short-lived; very soon two new developments undermined UAC's position. Firstly, in 1895, Hamilton Castner and Karl Kellner built a brine electrolysis plant in Runcorn, which provided a cheap source of chlorine for the Brunner

Mond company. Secondly, during World War I the company was required by the government to make ammonium nitrate and other intermediates for high explosives and to employ the Castner–Kellner chlorine plant to produce phosgene (carbonyl dichloride) and mustard gas (dichloro-diethyl sulfide) as chemical warfare agents. As a result, Brunner Mond was transformed into a broad chemical enterprise, which grew to become a founder member of Imperial Chemical Industries (ICI) in 1926 – for many years Britain's largest chemical company.

Useful commodities

Sulfuric acid remains the largest-volume chemical commodity globally, its annual output approaching 300 million tonnes. Around half of this sulfuric acid is used in agricultural chemicals, especially to produce fertilizers, such as ammonium sulfate and superphosphate of lime (mixture of calcium dihydrogen phosphate and calcium sulfate). These replace the nutrients that crops remove from the soil in order that farmers can produce highly nutritional food for feeding growing global populations.

The total global production of soda ash is currently estimated to be about 65 million tonnes. Over 75% of the world's sodium carbonate is still made using the *Solvay process;* the remainder is derived from natural deposits. Sodium carbonate is used as a cleansing agent and as a component of many soap powders. In glassmaking it serves as a flux for silica, lowering the melting point. It is used in treating water for removing calcium and magnesium ions that cause water hardness and as a pH regulator in swimming pools and aquariums. In toothpastes it acts as a foaming agent and an abrasive and it removes acidity.

Sodium carbonate is widely employed in the food industry – as a food additive, acidity regulator, anticaking agent, raising agent and stabilizer and when preparing ramen noodles. They remain truly ubiquitous commodities.

B: Bakelite and plastics

Black Bakelite Ekco Wells-Coats radio (1934)

Material of a thousand uses

Bakelite, the world's first fully synthetic plastic, was invented by pioneering chemist Leo Baekeland. The word 'plastic' is derived from the Greek *plastikos* meaning mouldable. Bakelite could be extruded, moulded into any shape and coloured. It was an electrical insulator and heat-resistant and was used to make many household items, billiard balls, radio cabinets and electrical components. It opened the door to the 'plastics age' and initiated the growth of a worldwide plastics industry based on a variety of synthetic polymers. Plastics are adaptable, lightweight, durable, flexible and cost-effective. They have made consumer goods more widely accessible and enabled critical developments in innumerable fields to improve society for the better.

History

History has been shaped by the materials used by people – from primitive buildings in wood, using stone to make tools, producing alloys in the Bronze Age to smelting iron during the Iron Age. Ancient people used natural plastics based on tree rubber, animal horn and shells. In the 20th century synthetic plastics became essential as hundreds of familiar products were made (in part) with plastics: car components, electronic devices, computers, phones, sports equipment, medical supplies such as joint replacements or surgical gowns and tools, contact lenses, toys, carpets, household appliances, building materials, signage, office supplies, packaging and synthetic fibres.

Several early pioneers of chemical technology modified natural materials and colloids (mixture of one substance of dispersed insoluble particles suspended throughout another) to produce new plastic materials. In 1839 Charles Goodyear in the US developed vulcanized rubber, a durable cross-linked polymer, used in tyres (see T). Then Swiss chemist Georges Audemars made the first synthetic silk fibres using mulberry bark and rubber gum. British chemist and inventor Alexander Parkes (1813–1890) made a synthetic plastic called *Parkesine* in 1862. In 1871 the US inventor John Wesley Hyatt (1837–1920) chemically modified cellulose to produce an astonishing new product called celluloid. This was made by applying heat and pressure to a mix of cellulose nitrate and camphor.

Hyatt was seeking to find a replacement for tusk ivory used for billiard balls, which had become scarce because of elephant hunting. Celluloid not only resembled ivory but had useful properties: at normal temperatures, it was a permanent, hard solid; when heated, it became soft and could be moulded or rolled into sheets. It soon became suitable for not only billiard balls but other products such as hair combs, buttons, shirt collars and then cinema film. Many other synthetic materials based on cellulose followed, such as viscose (cellulose xanthate). This artificial silk fibre was made by dissolving cellulose in carbon disulfide and cellulose acetate. The plastic explosive gelignite was developed by Alfred Nobel based on nitrocellulose, while in 1885 George Eastman of Kodak revolutionized photography by making a plastic photographic film from cellulose.

Another 40 years were to pass before the invention of the first wholly synthetic thermoplastic by Leo Baekeland, which expanded its use into mass-market consumer goods.

Pioneer

Leo Baekeland (1863–1944) was born in Ghent, Belgium. He came from a very modest background – his father was a shoemaker. He developed an interest in chemistry from his love of photography. He was awarded a scholarship to study chemistry at the University of Ghent and awarded his PhD at the age of just 21. By 1887 the university had appointed Baekeland as assistant professor of chemistry at the start of his promising academic career, although he soon became less interested in pure chemistry than in its potential applications.

While on his honeymoon in 1889, as part of a travel scholarship to New York, he was persuaded to stay in the US and apply his talents to solving chemical problems for industry. Baekeland took a job as a chemist with a photographic supplies house in New York. There he worked on improving photographic papers, which at the time could only be developed in bright sunlight. He eventually came up with a novel type of paper, Velox, that could be developed by gaslight. The new Velox paper soon saw booming sales, which quickly attracted the attention of George Eastman, founder of the Eastman Kodak company, who bought the invention for $750,000 – a huge amount for the time. Baekeland spent a share of the proceeds on the purchase of a luxurious new house incorporating a modern chemistry laboratory, which allowed him to pursue whatever scientific opportunities he wished, in particular the search for new synthetic materials.

Baekeland had set himself the goal of finding a synthetic alternative to shellac, the natural product made from the resinous secretions of female lac beetles (*Kerria lacca*) found on certain trees in east Asia. It was used to produce moulded consumer goods such as inkwells, jewellery, phonograph records and picture frames or used as a durable vanish for wood. Natural shellac was expensive and laborious to produce, so the potential demand for a synthetic alternative that could be mass-produced was considerable. Knowing that, in 1872, the German chemist Adolf von Baeyer had reacted phenol and formaldehyde to form a hard resin, Baekeland, in the summer of 1907, changed his focus from trying to create a wood coating to trying to strengthen wood by impregnating it with a similar synthetic resin. Then he turned his attention to a replacement for natural rubber as a binder for asbestos. In one promising experiment he boiled formaldehyde and phenol to obtain a viscous liquid, which was then heated in a sealed tube filled with asbestos fibre for four hours at 140 °C.

A day later, Baekeland had obtained four different condensation products, one of which was an amber-yellowish, hard substance, a thermosetting plastic that would not soften or dissolve in any solvents. Baekeland wrote: *'This looks promising and it will be worthwhile to determine how far this mass is able to make moulded materials either alone or in conjunctions with other solid materials, for instance asbestos, casein, zinc oxide, starch, different inorganic powders and lamp black and thus make a substitute for celluloid and for hard rubber.'* He called his new product 'Bakelite'.

'Bakelizer' steam pressure vessel used by Leo Baekeland

Baekeland had subjected the reactants to heat and pressure using a sealed autoclave or pressure vessel that he called the 'Bakelizer'. He confirmed that the sticky, amber-coloured resin would, when mixed with fillers, yield a hard-setting mouldable plastic that was not as brittle as earlier efforts, and the production process was fast and easy to control by manipulating the temperature and pressure. As a thermosetting resin, it would not lose its shape even when subjected to considerable heat or various solvents and it was a good electrical insulator. Baekeland took out many patents related to the manufacture and applications of Bakelite.

In 1910 he started semi-commercial production in his laboratory producing a daily output of 180 litres, He then formed a company to manufacture and market his new industrial material. By 1930 the Bakelite Corporation had grown to occupy a 128-acre plant at Bound Brook, New Jersey, and Bakelite plants were also set up in Britain and Germany. The material was easier to mould and cheaper to produce than celluloid. It was a simple process, which made it ideally suited to mass production and it could be coloured. It was marketed as *'the material of a thousand uses'*, as it found application in jewellery, buttons, ashtrays, handles, car distributor caps, adhesives and, famously, electrical insulators, early radio sets, telephones and light bulb sockets. It was rather brittle and needed to be strengthened with filler substances, making the colour rather dull.

In 1939 Baekeland became very wealthy from selling his company to Union Carbide. By 1944 global annual production of Bakelite had reached 175,000 tonnes and it was found in over 15,000 different products. Bakelite use declined with the arrival of many other new synthetic polymers with a wider range of properties – but it was Baekeland's innovation that had initiated the modern plastics industry.

Bottles made from PET- a clear lightweight plastic

Plastics for everything

In 1930 the American chemist Wallace Carothers at DuPont discovered nylon, which was to revolutionize textile manufacture, and the German chemist Eduard Simon made polystyrene. Waldo Semon (1898–1999), a young chemical engineer with the B.F. Goodrich Company in Ohio, attempted to invent a method for converting a waste vinyl chloride into an adhesive. Although Semon failed, when heating the material he inadvertently discovered a polymer that was both flexible and elastic. In 1933 Semon patented the process to produce polyvinyl chloride (PVC) from polymerization of the vinyl chloride monomer. The new product had a variety of uses including waterproofing and flooring, plastic piping, windows and of course the manufacture of 'vinyl' records.

Another ground-breaking new plastic had first been synthesized in 1898 by accident by German chemist Hans von Pechmann, who prepared a white waxy polymer by heating unstable diazomethane. The British chemists Reginald Gibson and Eric Fawcett rediscovered it while undertaking research investigating high-pressure reactions at ICI in Winnington, UK. In 1933, following an experiment that had subjected a mixture of ethylene and benzaldehyde to a pressure of 1,900 atmospheres and a temperature of 170 °C, Gibson and Fawcett noted an unexpected pressure drop and quantities of a waxy solid in the reaction vessel. This was later found to be polymerized ethylene. It was to become the most widely used of all plastics. A member of their team said:

> *We'd been heating it … to well over boiling point. For once it did not explode – (usually it did!) – and we thought something must be wrong. So, we left it to cool overnight. And when I looked inside the metal container the next day, I found what looked like a lump of sugar – but it was… polyethylene.*

It was believed at the time that ethylene could not polymerize because its double bond could only be activated at very high temperatures, but in this trial the reaction had been initiated by trace oxygen contaminating their apparatus. The experiment proved difficult to reproduce, was often explosive and because of safety concerns they halted their research. The research team later discovered that the polymerization required a specific amount of oxygen to initiate. At the time, it was common practice to return gas cylinders for refilling with their valves open. This meant that the cylinders could contain anywhere up to 1 atmosphere of air, leading to ethylene refills with a variable quantity of oxygen. The oxygen mixed with the ethylene had a huge impact on the reaction: if completely absent, there was no reaction; at 0.002% the reaction was stable; but if the concentration reached 2% or more, the mixture exploded. Only by chance had Gibson and Fawcett picked ethylene canisters containing just the right level of residual oxygen to act as an initiator. Understanding the role played by oxygen provided vital in their research, which soon started to yield much more predictable and interesting results. ICI realized that it had stumbled upon a material with considerable potential. Later in December 1935 another chemist in the team, Michael Perrin, developed a reproducible high-pressure synthesis for polyethylene that became the basis for industrial low-density polyethylene (LDPE) production.

High-pressure chemistry was still a new area of manufacturing during the 1930s as designing and building strong reactors was not straightforward. In July 1939 the pilot plant was upgraded with the first of two 50-litre reaction vessels. The larger-scale 100-ton-per-annum plant was moved into full-scale production on 1st September 1939 – the day that Britain declared war on Germany.

There was soon much interest in using polyethylene from telecommunications companies, who needed to insulate transatlantic cables. It was also used in insulating airborne radar, which gave a critical advantage in the Battle of the Atlantic by helping British supply ships avoid German submarines. These early applications for polyethylene made use of its high dielectric strength, low loss factor, moisture resistance, flexibility and lighter weight than other insulating materials. This enabled Britain to create radar systems that were light enough for fighter planes. During the war the production of the plastic was secret but afterwards it became a versatile commercial product.

In 1944 ICI licensed both DuPont at Sabine River, Texas, and Bakelite Corporation at Charleston, West Virginia, to begin large-scale commercial polyethylene production. The real breakthrough in polyethylene manufacturing began with the development of catalysts that allowed polymerization at lower temperatures and pressures. The first catalyst was based on chromium trioxide jointly discovered by Robert L. Banks (1921–1989) and Paul Hogan (1919–2012) at Phillips Petroleum Company, Bartlesville, Oklahoma, in 1951.

The problem limiting the wider use of LDPE was that it was notoriously soft and had a low melting point, which made it unsuitable for any applications involving heat. In the high-pressure polymerization process, the ethylene molecules did not always add together in a regular chain. This problem was overcome by German chemist Karl Ziegler, who developed catalysts that resulted in a low-branching, straight-chained polymer that was much more rigid and could handle boiling water. In 1953 he developed a catalytic system based on titanium halides and organoaluminium compounds that worked at even milder conditions with lower pressures than the Phillips catalyst. Although the Phillips catalyst was less expensive than Ziegler's, it was also employed industrially to produce the new successful high-density polyethylene (HDPE). Hogan and Banks improved the process so it only required lower pressures and produced HDPE that was far stiffer, harder and more heat-resistant than its competitors. The new discovery launched Phillips into an entirely new industry manufacturing a family of synthetic plastics. They introduced HDPE in 1954 under the brand name *Marlex* polyethylene although it was still unsuitable for some applications. The early HDPE was too brittle and prone to cracking after a few months. To use up their large stockpiles of brittle HDPE they decided to make large rings of plastic tubing to be sold as 'hula hoops', the popular children's toy and adult fitness aid.

Giulio Natta, an Italian chemical engineer, developed further variations of the Ziegler catalyst based on his findings on the mechanism of the polymerization reaction. The new Ziegler–Natta catalysts included mixtures of halides of transition metals, especially chromium, titanium, vanadium and zirconium, with organic derivatives of non-transition metals, particularly alkyl aluminium compounds. In 1963 both men were awarded a Nobel Prize for their pioneering work on catalysts.

Today, polyethylene is considered one of the world's most versatile and widely produced plastics. Global production exceeded over 100 million tonnes in 2020. Being a true thermoplastic, it can be injection moulded, extruded or blow moulded. LDPE is used in freezer bags, bubble wrap and food containers, detergent bottles, margarine tubs and plastic toys. HDPE is used to make such items as medical ware, fuel tanks, insulation, seating, binders and furniture.

Chemical technology proved critical for the Allied victory in World War II. Americans searched for alternative materials that would substitute for or improve the performance of scarce natural resources like rubber, metal, wool, wood and cotton. Apart from the success of using polyethylene in electronics, the advances in polymer science provided many new plastics that proved invaluable to the war effort. For example, nylon fibre was used for parachutes, ropes and netting. *Plexiglas*, a solid, transparent plastic made of polymethyl methacrylate, provided a light, durable alternative to glass for aircraft windows. When the war ended, the American economy had shifted to using the new plastics to produce a vast range of consumer goods.

In 1938 a DuPont chemist, Roy Plunkett, was experimenting with new refrigerants when he found his canister coated with a slippery white substance. This was the polymer polytetrafluoroethylene (PTFE), which he recognized had frictionless qualities and could have many useful applications. It was marketed as *Teflon*, for use in everything from medical equipment to non-stick pans and later used in breathable clothing as *Gore-Tex*. In 1966 Stephanie Kwolek and Paul Morgan of DuPont produced a heat-resistant aromatic polyamide synthetic fibre with a molecular structure of many inter-chain bonds. This makes it incredibly strong and super-tough but it can be woven like nylon. These chains are cross-linked with hydrogen bonds, providing a tensile strength 10 times greater than steel on an equal weight basis. This was *Kevlar*, best known for its use in ballistic body armour, but it has many applications because of its high tensile strength-to-weight ratio.

In 1987 Charles Hull, an American furniture builder who was frustrated with not being able to easily create small custom parts, developed a system for creating 3D models by curing photosensitive resin layer by layer. Using his process called '*stereolithography*', where layers of photopolymers are cured with ultraviolet rays to form solid objects, he produced the first commercial 3D printer that could print a real physical item from a digital computer-generated file. Today, rapid prototyping using 3D printers has become an essential technology in the medical, automotive and aircraft industries. One of the plastics employed in 3D printers is the thermoplastic ABS (acrylonitrile butadiene styrene), which is the material from which LEGO bricks are made.

Plastics for the future

Modern standards of living have improved dramatically due to the benefits of plastics, whose low cost and versatility have made a wide variety of goods more accessible. Plastics have made possible the vast revolution that spread digital electronics, mobile phones and small high-powered computers. Plastics have improved safety, with inventions like padded foam dashboards, polycarbonate glass and bicycle helmets, while greater fuel efficiency has been possible making use of plastics to produce lighter cars and aircraft. The use of plastics in medicine has led to better hygiene, advanced treatments and artificial limbs. The global plastic market is estimated to be over $600 billion.

It has recently been acknowledged that the dominance of plastics may cause environmental problems due to their slow decomposition rate in ecosystems and that they can lead to marine pollution unless carefully controlled in supply chains. Their production requires the combustion of fossil fuels, which also impacts the climate. INEOS is building Europe's most energy-efficient, lowest carbon emitting and technologically-advanced ethane crackers (*Project One*) in Antwerp to produce ethylene and propylene for plastics. Some companies are developing new recycling processes and finding chemical technologies to make truly biodegradable plastics. New types of plastic are being made from renewable materials, such as plant starches and sugars. The chemical technology of plastics is still very much evolving. L'Oréal, the cosmetics company, is making product bottles from PET plastics entirely recycled using a new enzymatic technology.

It is certain, however, that in the future plastics and other synthetic polymers will continue to play a vital role in medicine, food, electronics, clothing, construction, aerospace and transport.

C: Cement

Building the modern world

Cement is the product of chemical technology used to bind sand and other materials together and is one of the most widely used construction materials. The modern built environment is unimaginable without the widespread use of cement-based materials that allow the construction of strong, complex and sometimes massive structures. Few buildings, factories, hospitals, roads, bridges, water reservoirs, sewers, seaports or airports could exist without the use of cement. Cement is used for renders for facing buildings, mortars for bricklaying, building foundations, casting concrete blocks or beams, paving and roof tiles. Most construction cements are used in combination with aggregates – coarse to medium particulate materials such as sand, gravel, crushed stone, slag, crushed recycled concrete and geosynthetic materials. Over 50% of the cement used today is as part of the strong concrete poured into steel-reinforced concrete foundations and to form structural components.

Portland cement is the most common type as an ingredient of concrete and mortar. It is a hydraulic cement – one that not only hardens by reacting with water but also forms a water-resistant product. Originally it was a coarse powder, produced by heating limestone and clay minerals in a kiln to form clinker, grinding the clinker and adding small amounts of gypsum (calcium sulfate).

The early chemical technology of Portland cement was developed from a hydraulic lime by Joseph Aspdin (1778–1855). It was aimed at the market for renders (stuccos) and architectural pre-cast mouldings, for which a fast-setting, low-strength cement was required. However, it is really his son William Aspdin who can be regarded as the inventor of modern Portland cement because of improvements he made in the 1840s. It became available during a period of ambitious infrastructure building at the time of the Industrial Revolution.

History

Cement was used in Ancient Greece and Rome. The Romans used a mixture of lime (calcium oxide) and pozzolan (crushed volcanic ash) to create hydraulic cement, which could set under water. Some of the famous historical buildings made from concrete can still be seen, including the ruins of the Colosseum and Pantheon in Rome and the Hagia Sophia mosque in Istanbul. In the Middle Ages masons used hydraulic cements to build large structures such as castles, cathedrals and canals.

The Industrial Revolution that started in Britain in the late 18th century saw a flurry of developments in cement and concrete manufacturing. An important contribution was made by John Smeaton, who discovered that the property of a lime binder to harden in contact with water (hydraulicity) was directly related to the limestone's clay content. In 1756 he experimented with combinations of different limestones and additives, including volcanic trass and pozzolan, while working on lighthouse construction. A cement was developed in 1796 by James Parker called 'Roman' cement, and in 1822 James Frost at Harwich patented and made an artificial cement he called 'British' cement. Edgar Dobbs of Southwark, London, patented an artificial hydraulic lime as a forerunner of Portland cement. The true precursor to modern Portland cement was created by Joseph Aspdin in 1824.

Profile

Joseph Aspdin was the son of a bricklayer living in Leeds, Yorkshire. He entered his father's trade and had by 1817 set up in business on his own. He experimented with cement manufacture by heating limestone (calcium carbonate) and clay until the mixture calcined (lost carbon through combustion) to yield quicklime (calcium oxide) and carbon dioxide. In October 1824 he was granted a patent for his cement (British Patent BP 5022 entitled *'An Improvement in the Mode of Producing an Artificial Stone'*). The process was described as 'double burning' in which the limestone was burned, then 'slaked' by combining it with water, then mixed with clay and burned again. The grinding technology of the time consisted only of flat millstones, so it was more economical to pulverize the limestone by burning and slaking than by grinding.

He named it Portland cement after its resemblance to the famously strong building stone (Oolitic limestone) quarried on the Isle of Portland in Dorset, England. At the time Portland stone was the most prestigious building stone available in Britain. His new artificial cement was developed to compete with Parker's Roman cement.

Joseph made use of waste pieces from limestone road construction as a cheap source of raw material, but he often ran short of stone and was prosecuted for digging up whole paving blocks from local roads. This early cement was not the same as modern Portland cement, but it was an important step in its development. It was fired at relatively low temperature (below 1,250 °C) and contained no alite mineral (an impure form of tricalcium silicate).

Joseph's younger son, William Aspdin, was running the family cement plant in Leeds but after a financial dispute with his father he established his own plant at Rotherhithe near London in 1843.

There, William created a new and substantially stronger cement of his own recipe. However, he did not file for a new patent on his modified process and relied on his father's earlier patent. His innovation was to make a mix with a higher limestone content, to burn it at a higher temperature then to grind the hitherto discarded 'clinker'. He had realized the usefulness of these hard, overburnt rocks of the lime and clay mixture, which were previously considered useless in cement and thrown out. He found that the higher firing temperatures produced calcium silicates that, when mixed with water, gave the cured concrete much greater strength. He had made the first cement containing alite. William ground up the clinker, packed the powder in casks and sold it as his new cement. While lacking a scientific education he had none the less discovered the key step in the development of cement.

The new version of Portland cement had exceptional strength as when mixed with sand and water and allowed to dry it hardened into a concrete that was almost *twice* as strong as the existing cements. William found a financial backer and started advertising his product as an improved version of his father's Portland cement and it proved a commercial success. In 1848 William moved south to Northfleet in Kent, which has inexhaustible supplies of soft chalk. Soon his competitors wanted to reproduce his cement as his clinkering process had no patent protection, obliging him to be very secretive and personally supervise all the production.

In 1845 Isaac Johnson (1811–1911), a rival from London who had studied chemistry and managed a cement works in Swanscombe, undertook months of scientific experiments that led him to understand the value of clinker. He fired chalk and clay at much higher temperatures than Aspdin (1,400–1,500 °C), leading to the mixture clinkering and producing a modern-day version of Portland cement. Although Aspdin continued manufacturing but his firm eventually went bankrupt. He died in 1864 at the age of 48. Johnson eventually took over Aspdin's cement works and made it a financial success, including obtaining three patents for new cement. William Aspdin, although often forgotten, had an undeniable role in the development of the crucial chemical technology for building the world's infrastructure.

In 1859 John Grant of the Metropolitan Board of Works set out requirements for the cement to be used in the construction of the London sewerage system. This became the standard specification for Portland cement. The lime is obtained from a calcareous (lime-containing) raw material, and the other oxides are derived from clay-containing materials. Smaller quantities of additional raw materials such as silica sand, iron oxide and aluminium minerals may be used to get the desired composition.

The next advance in the manufacture of cement was the rotary cement kiln, patented by Frederick Ransome in 1885. The kiln allowed a uniform clinkering temperature, ensured a more homogeneous mixture and a continuous manufacturing process.

Cement factory New South Wales, Australia

The process for making Portland cement involves firstly the quarrying and physical extraction of raw materials such as limestone, shale, silica and iron oxides. These are transported to the cement plant to be crushed and milled into fine powders. These powders are blended into the 'raw meal' and preheated to around 900 °C using the hot gases from the kiln to burn off the impurities. The material is then heated in large rotary kilns at 1,500 °C. Heating starts the decarbonation or calcination in which carbon dioxide is driven from the limestone.

The partially fused product that results is clinker. A modern kiln can produce around 6,000 tonnes of clinker a day. The clinker is then cooled and ground to a fine powder in a ball mill. This mill has a rotating drum filled with steel balls of different sizes (depending on the desired fineness of the cement). Gypsum (calcium sulfate dihydrate) is often added during the grinding process to control the setting rate of the cement. Finally, the cement is bagged off and transported to customers. Cement must be handled carefully as, being alkaline, it can cause chemical burns, while the fine powder can cause irritation or, with severe exposure, lung diseases as it contains hazardous mineral components.

By the 1880s many Portland cement plants had been built around the world. The low cost and widespread availability of the limestone, shales and other naturally occurring materials made cement one of the lowest-cost and most widely used materials and it became one of the most versatile construction materials. Concrete is made by combining water with cement to form a cement paste. The paste glues the added aggregates together, fills voids within it and makes it flow freely. A lower water-to-cement ratio yields a stronger, more durable concrete, whereas more water gives a freer-flowing concrete. The best mix is a compromise between strength and workability. Portland cement contains compounds of calcium silicates and aluminates, which all undergo hydration to contribute to the final material's strength. With time, the products of the cement hydration process gradually bond together the individual sand and gravel particles to form a solid mass. The process is considered irreversible. Concrete cures in several stages and the mix eventually hardens. Most of the hydration process is complete after about a day, but the cement will continue to cure for a long time while water and unhydrated compounds are present. This can take some months but its strength is normally tested after 28 days. There have been some massive concrete structures built, such as the Grande Dixence Dam in Switzerland, the world's tallest dam, in which 6 million cubic metres of concrete was used in its construction in the 1950s.

Concrete growth

Driven by massive industrialization and increasing urbanization, the building of cities with houses, hospitals, dams, roads, bridges, railways and service infrastructures has increased in scale over the past half century, with an ever-growing demand for cement-based materials. They are at the heart of the modern built environment and use more than one-third of the total materials extracted from the earth each year. Since 1950 global population has increased about 3-fold, but the amount of cement produced has increased over 30-fold as living standards have risen in most parts of the world. Cement production is one of the largest industries globally, reaching over 4 billion tonnes in 2023. The fast-growing economies of countries such as China and India now produce well over half of the world's output of cement. As global economic and population growth continue, demand is being fuelled to produce even greater quantities of cement.

As cement production is an energy-intensive process with energy costs constituting more than 60% of its total cost, modern cement plants must maximize energy efficiency. The environmental impacts from cement not only include the high energy for manufacturing because of the high temperatures employed but also the energy required to mine and transport raw materials. Cement production generates more carbon emissions than any other industrial process, both from direct carbon dioxide emissions associated with combustion of the fuel in the kilns but also from the chemical reaction used to produce the clinker and from thermal decarbonation (calcination) of limestone (calcium carbonate) to calcium oxides. The emissions from cement production, coupled with the high rate of consumption, make concrete responsible for 5–8% of global carbon dioxide emissions (1 tonne of cement produces about 0.8 tonnes of carbon dioxide). Poorly controlled plants can also give rise to air pollution, potentially releasing dioxins, nitrous oxides, sulfur dioxide and particulates. This clearly illustrates the dilemma between the benefits of global growth and sustainability and environmental protection.

Although cement manufacturing is a large greenhouse gas emitter, over its long lifetime it compares quite favourably with other building systems – such as wood, steel, asphalt and brick, none of which could be sufficiently increased to replace concrete to any significant extent or scale. New cements are also being developed that can absorb carbon dioxide over their lifetimes. One process involves 'carbon sequestration' as the hydrated cement reacts with carbon dioxide in the air, slowly reversing some of the emissions that took place in the kiln when the cement was made and reducing the carbon footprint of the building material. The use of low-carbon supplements, such as a graphene additive (see K), to increase strength or reducing the clinker level in cements could both help to reduce the carbon footprint and potentially increase longevity of products. Carbon capture technologies will also have an important role in improving the sustainability of cement and concrete. Researchers in Cambridge have recently invented a process using recycled cement to produce ultra-low-emission concrete while also cutting the carbon footprint of steelmaking. Also available is *Cemfree*, a proprietary cementitious material that can make concretes with embodied carbon dioxide that is up to 80% lower than Portland concretes.

D: Dyes

Bringing colour to life

Dyes are chemicals that add colour and vibrancy to the world around us. While they occur in nature, they are produced by many chemical technologies in the dye and pigment industry. These colour an endless array of clothing and fashion goods, home furnishings, leather, paper, inks, liquids, plastics, electronic displays, decorative coatings, building and construction materials, glass and even foodstuffs.

The first truly synthetic dye was made in 1856 by a British chemistry student called William Henry Perkin, who discovered a mauve-purple dye made from aniline derived from coal tar. Perkin's discovery began a new organic chemical industry. He was at the time attempting to produce quinine to combat malaria but found he had serendipitously made a synthetic dye. Perkin had seen that one of the by-products of his reaction was purple, and he went on to ascertain that it might be used for colouring wool and silk. He patented this first synthetic dye and began to make plans to manufacture commercially at scale. He developed a process for the production and application of the new dye. In 1857 he opened his new dye factory, which made him wealthy but also drove up interest among the chemists of the day. He was not slow to exploit his chance discovery of the first truly synthetic dyestuff and thus established a great new industry. It was said of Perkin that *'his enduring fame ... rests on the fact that he was the first to unlock the limitless treasure chest of man-made colour, and by this to bring new beauty into the lives of millions all over the world'*. This industry also stimulated the search for a better understanding of the chemistry of organic compounds, leading to new products such as medicines and pesticides.

Over 60% of dyes made today are used for dyeing textiles. These dyes need good colour fastness to impart colour, design and feeling to the fabrics. The dyes absorb and reflect light at the different wavelengths that give the human eye a sense of an object's colour. Dyes have chromophore and auxochrome in their molecular structure, which selectively absorb visible light differently and therefore produce the different colours or hues. The degree of their brightness depends on the reflectance of the dye molecules to incident light and its saturation is related to purity of the colour.

History

The procuring of fast and beautiful colours for the adornment of people and material has been an absorbing interest from earliest recorded times. Traditionally dyes were soluble compounds made from plant extracts, fixed to fibres directly or requiring a 'mordant', such as a metal salt, to ensure the colour remained after washing. Pigments were largely insoluble particles produced from minerals and earths suspended in a medium such as an oil or fat to make paints or inks.

There is archaeological evidence for the use of dyestuffs and pigments in ancient times. There are cave paintings in France using yellow and red pigments from ochre and carbon black dating from 15000 BC. The seeds of weld plants (*Reseda luteola*) used to make a yellow dye were found in Neolithic excavations dating from 6000 BC. There are Egyptian paintings using a blue ink and pigments based on crushed minerals such as red cinnabar (mercury sulfide) dating from over 3,000 years ago. Vermillion is an orange-red pigment developed in China and later common in Greece. The earliest written record of dyeing silk appears around 2600 BC in China. Woollen dyes were in use in Rome around 715 BC. Alexander the Great mentions the valuable purple robes he had seen in Susa, the ancient Persian capital (modern Iran), around 330 BC, and beautiful printed cottons were also known in India around the same time.

The Romans created a wide range of pigments for decorating wall murals. Their emperors used the *Murex* mollusc to create a purple dye costing its weight in gold, and no one apart from the Imperial family was allowed to wear it. It took some 10,000 molluscs to produce a mere gram of dye. In Britain and Gaul, invading Romans found native peoples using a blue dye made from the leaves of woad plant (*Isatis tinctoria*), which was cultivated to dye fabrics and themselves. By the Middle Ages, most dyes were based on vegetable sources, although they were far from water or light fast. In the less wealthy parts of the country, lichens, mosses and native plants were frequently used for many centuries to produce dyes for traditional clothing, including tweeds and tartans. Dyeing methods and recipes were jealously guarded, to be handed down in families from one generation to the next.

In 1464 Pope Paul II introduced 'Cardinal purple', a luxury scarlet dye made from a shield louse insect (*Kermes ilicis*). In Britain, as the country grew in wealth, there was a desire for rich, colourful tapestries, banners and silks. Elizabethan dress used newly discovered colours from the New World employing tinted woods such as brazilwood (*Caesalpinia sappan*) and carmine red, taken from female cochineal insects (*Dactylopius coccus*). Until the 19th century all dyes were produced from a wide range of such natural sources, whether plant or animal matter, insects, metals or earths. In 1856 over 75,000 tonnes of natural dye materials were imported to the UK.

The early alchemists, often surrounded by myth and secrecy, experimented with creating new coloured materials. Only in the 17th century did scientific progress tend to reduce the role of the alchemists, although they still made some remarkable discoveries in the field of dyestuffs. In 1704 a paint manufacturer named Johann Diesbach and a pharmacist Johann Conrad Dippel (1673–1734) in Berlin accidentally discovered a dark blue, made from cyanide, potassium and iron, with exceptional colouring strength – 'Prussian Blue'. The scientists of the 18th century began to develop a scientific understanding of chemistry.

An orange-red mineral called Siberian red lead (crocoite) was discovered in the Ural Mountains for use as a pigment by artists. In 1797 French chemist Louis Nicolas Vauquelin (1763–1829) analysed it. He detected a new metal, which he isolated and named 'chromium' from the Greek for colour, due to its extraordinary capacity to produce salts in many colours, including a fine bright yellow pigment of lead chromate and the brilliant emerald colour called 'Viridian green'. Soon the manufacture of chrome pigments proliferated throughout Europe. The bright new colours were produced on an industrial scale for paints to be used on walls, floors and furniture.

The chemists now sought to extract, isolate and characterize the colouring ingredients obtained from natural materials. In 1826 Pierre Jean Robiquet (1780–1840) isolated alizarin from madder and orcein, a red-violet dye obtained from lichens. As they began to succeed, the possibilities offered by chemical synthesis began to dominate thinking. Picric acid or 2,4,6-trinitrophenol (TNP), although later employed as an explosive, was used as a yellow silk dye. The large-scale agricultural production of dye crops would finally give way to the manufacture of synthetic pigments and dyes.

When industrial coal gas production was established in the early 1820s to produce fuel for lamps and heating, chemists gained a source of raw materials originating from coal tar distillation, which would be employed in the early synthetic colour industry. In 1826 German chemist Otto Unverdorben (1806–1873) extracted a pure substance when heating indigo and named it aniline (Arabic: *anil* for indigo). Most famously, aniline would be used to produce the first truly artificial dye and was perhaps the most commercially significant organic chemical discovery of the Victorian age.

The by-products of the production of gas and coke from coal were subjected to further distillation, which provided a range of marketable fractions such as the profitable aromatic products of benzene, aniline, toluene, naphthalene, anthracene and phenol or carbolic acid. In 1846 August Hofmann of the Royal College of Chemistry in London suggested that these aromatic compounds could form the basis for an organic chemical industry and asked that his student William Perkin synthesize quinine from naphthalene. At the time quinine was obtained solely from cinchona tree bark grown on plantations in Southeast Asia, and it was much in demand as the only known means of treating malaria.

Pioneer

William Perkin (1838–1907) was born in London and had begun his higher education at the Royal College of Chemistry in London. It was an early fascination for crystals that had first excited the young Perkin's interest in chemistry.

Organic chemistry, involving the carbon compounds that are the constituents of all animal and plant life, was only beginning to be understood. He had set up a simple laboratory in his father's house in the East End of London, where he continued to work on quinine during the Easter vacation of 1856. To this end, he heated a mixture of aniline sulfate and potassium dichromate. What he obtained was not quinine but a black and apparently unpromising precipitate. Perkin, being inquisitive, did not throw this away but found by adding alcohol it gave the substance a strong purple hue. Remarkably, unaided and at the age of only 18, he had discovered the world's first synthetic dye in his crude laboratory. His persistence led him to borrow a piece of silk from his sister and find that his new product dyed it an attractive purple shade. The complex chemical structure of his new mauve dye was first postulated in 1888 but it was only correctly determined many years later.

Perkin's success in founding a great industry at least equals his scientific achievement. He quickly saw the potential of the new material as a dye and tried it out on other samples of silk. Professor Hofmann tried to dissuade Perkin from leaving a promising academic career for what many Victorians regarded as the dirty and socially inferior world of industry. The manager, Robert Pullar, of Scottish commercial dyers in Perth encouragingly said that if this colour could be made cheaply enough, and if it was fast on silk and on cotton, *'it is decidedly one of the most valuable discoveries that has come out for a very long time ... the shades are certainly the best I ever saw; from what I have seen there appears to be no doubt of its success'*.

In the mid-Victorian era, purple fabrics were the height of fashion, being a favourite colour of Queen Victoria. Purple dye had traditionally been made from various kinds of shellfish since the days of Troy, but, because of its scarcity and cost, it became associated with the clothing of emperors, kings and aristocrats – hence the saying *born to the purple* indicated somebody of wealth and status who could afford a fabric dyed purple. Unfortunately, silks dyed by natural purple also quickly faded so that its use was flamboyant; further evidence of wealth and standing. Perkin obtained his dye patent on 26th August 1856 then conducted successful dyeing trials with Thomas Keith, the largest silk dyer in London. His father backed him with money, and William's brother Thomas also joined the business (Perkin & Sons).

Perkin pressed ahead, firstly developing his process for a larger scale, then searching for a suitable site. In mid-1857 he found a six-acre site on which to build his new dye-making factory at Greenford Green in the countryside of Middlesex. For transport, the factory would be near to the Great Junction Canal. Most of the process required new equipment to be purpose made, and he designed a cast-iron vessel holding 40 gallons with a stirring tool at one end and the lid fastened by a cross bar at the other. Funnels were built into the tops for charging acid and benzene, with a vent to emit nitrous fumes.

The manufacturing process took two days, combining aniline, sulfuric acid and bichromate of potassium, which produced a black solution that filtered to a soot-black powder. This was a mixture of the desired mauve dye and a range of by-products, which had to be removed by solvent extraction with naphtha and methylated spirits. Again, new apparatus was designed by Perkin for the extraction. Finally, the spirit was distilled off from the product, which was filtered and washed with caustic solution, leaving the dark mauve paste. Perkin said that from some 11 pounds (5 kg) of coal tars could be obtained 2¼ ounces (65 g) of aniline and only a ¼ ounce (7 g) of pure mauve dye. One pound (0.45 kg) of his purple dye could colour some 200 pounds (90 kg) of cotton fabric, although early attempts to apply synthetic dyes required Perkin to develop new dyeing methods and new fixatives. Suppliers were incentivized to find new ways of making the materials needed. His energy was astonishing, and by December 1857, less than two years from its discovery, the new dye was being delivered to its first customers and proved very popular.

Apart from his chemical discovery, his great achievement was the way in which Perkin planned, designed and built his first factory, including the design of suitable process equipment – an early instance of chemical engineering design through empirical studies and pilot trials. He can be considered as both an applied chemist and a prototype for a modern chemical engineer. From this modest beginning grew the highly innovative synthetic dyestuffs industry and many associated chemical enterprises, although he is often forgotten today.

The new mauve dye found early success in France, as the Empress Eugenie (the wife of Napoleon III) was one of the first to wear it. The colour was like the mallow flower, which led to the idea of using *Mauveine* as the trade name. Perkin's Mauveine would transform the market, making this prized colour readily available, whether for clothing or even postage stamps. What has been described as 'an overwhelming mania' for the new dye spread throughout Britain by 1859. The Greenford works prospered, more new plant was erected and further suppliers of aniline had to be found. As ever, fashion moved on so that the production of Mauveine reached its peak three years later. Perkin's invention had by then inspired others to create more synthetic dyes based on aniline.

William Perkin holding dyed silk (1907)

Synthetic dyes grow

After Mauveine, William Perkin went on to develop a practical manufacturing route to synthetic red alizarin from anthracene, which was perhaps of even greater industrial and commercial significance. Natural alizarin, obtained from the madder root, had been used for centuries to make a prominent red dye. Two German chemists Carl Gräbe and Carl Liebermann, working at BASF also isolated alizarin. They had separately applied for a patent for synthetic alizarin in June 1869 and so joined forces with Perkin to market it. By 1873 annual capacity of the alizarin plant was about 450 tonnes, with his company making a profit of around £60,000 (over £6 million at today's value). Synthetic alizarin production replaced the traditional cultivation of natural madder in Holland and France in less than 15 years. The discovery of a synthetic alternative to the widely used natural blue indigo dye by Adolf von Baeyer in 1887 led BASF eventually to develop a commercial manufacturing process that replaced the indigo produced from plant materials in India.

Perkin rather fell victim to his own success with alizarin, as he needed room for further expansion of his works and workforce, a new water supply and suitable drainage, and he had also become concerned about the numbers of accidents and losses in production. Nor was his patent protection watertight so at the age of 36 he sold up the business and retired to carry out research in his laboratory. He even made vanilla-scented coumarin, the first fragrance to be made artificially. Perkin can justifiably lay claim to the title of 'father of the organic chemicals industry'. He was one of the remarkable Victorian innovative entrepreneurs with a thirst for knowledge, a genius for invention and unstoppable business drive. After the first synthetic dye, organic chemistry had become central to the modern age – exciting and profitable, replacing the traditional trade based on natural and plant-derived materials. British coal tar production was suddenly to rise from 175,000 tonnes in 1870 to 640,000 tonnes within 10 years, a large part of which was distilled into the key intermediates of dye manufacture. Effectively, it was the disposal of the waste products of the coal-based process generating town gas, road tars, pitch and creosote that went on to provide many more products of industrial importance into the early 20th century. Until the arrival of petrochemicals much later, the manufacture of all the products that were developed over the subsequent 70 years – dyestuffs, plastics, pharmaceuticals and so on – were largely based on the by-products of coal distillation. Britain would not be able to retain its early lead in this technology and was beaten by fierce competition from its better organized German rivals. When World War I broke out in 1914, Britain was importing three-quarters of its dyes from Germany. This left a serious shortfall.

British entrepreneurs such as James Morton, the managing director of Morton Sundour Fabrics, was forced in 1914 to start producing his own light- and wash-fast vat dyes. He founded a large factory in Grangemouth, Stirlingshire, and later sold his company, Scottish Dyes Ltd, enabling it to eventually become part of Britain's largest chemical company (ICI), which for many decades carried Perkin's colourful legacy forward. There have been many developments in colourants since the discovery of the first synthetic dye by Perkin. When new synthetic fibres such as polyester and nylon arrived, there was research into suitable new dyestuffs. The introduction of reactive dyes that formed a covalent bond with the fibre during the process of dyeing, had a high resistance to fading and were available in a range of bright shades made them suitable for colouring cotton and rayon.

Colouring the future

The combined global dyes and pigments market today is valued at over \$30 billion and is expected to grow as demand for textiles, paints and coatings, construction materials, electronics and plastics continues to increase. The increasing production of colourants results from the growth in global population. More recently, organic dyes have become important in electronics and optoelectronics. They can be used in optical data storage, solar cells and biomedical sensors.

New uses for dyes are today coupled with utilizing advanced application technologies, efficient manufacturing processes, reducing greenhouse gas emissions and adopting environmentally beneficial practices.

E: Electronic semiconductors

The silicon age

Electronic semiconductors are one of the most significant technological achievements to evolve from solid-state physics but they have been facilitated by chemical technology. They have made possible enormous advances in electronics, computers, smartphones, wireless communication, solar cells and space exploration and are now an essential part the modern digital and internet age.

The development of the semiconductor-based transistor led by William Shockley, an American physicist, was the first major advance. He was working at Bell Labs with two other scientists when they demonstrated the world's first successful transistor in 1947. When Bell Labs announced their breakthrough, they said that *'the transistor may have far-reaching significance in electronics and electrical communication'*. They were right. Unlike the vacuum tubes or thermionic valves of the time, transistors required very little power, generated less heat and required no time to warm up. Most importantly, they were to evolve into 'microchips' containing millions of transistors connected in integrated circuits. Transistors could perform millions of times more calculations occupying millions of times less space. It was partly because of Shockley's attempt to commercialize the new transistors in the Palo Alto valley near San Francisco, California, that 'Silicon Valley' became the foremost global centre of electronics and computing innovation.

While a substance that conducts an electrical current, such as the metal copper, is called a conductor and a substance with a very high resistance to current, such as glass or polythene, is called an insulator, semiconductors are materials with properties somewhere in between. The semiconductors, such as silicon and germanium, have four outer shell electrons. In a pure crystal of silicon each silicon atom is surrounded by four other atoms. In this state the silicon will not conduct a current unless it is 'doped' with tiny quantities of impurities into the molten material before it crystallizes to modify its electronic structure. The atoms of the dopant phosphorus, for example, may have five electrons in their outer shell, or in the case of the dopant boron, three atoms in their outer shell. The phosphorus atom in the silicon crystal lattice will form four bonds with adjacent silicon atoms, leaving one unbonded electron 'spare'. This electron will move in an applied electric field creating conduction. In the case of boron atoms there is a residual positive 'hole' that can move through the lattice.

The nature of semiconductors is explained in solid-state physics using a model describing the states of electrons known as band theory. Solid materials can have values of energy only within certain specific ranges. The behaviour of an electron in a solid (and hence its energy) is related to the behaviour of the other particles around it.

Pioneer

William Shockley (1910–1989) grew up in Palo Alto, California. He earned his Bachelor of Science degree from Caltech in 1932 and then a doctorate from the Massachusetts Institute of Technology (MIT) in solid-state physics. He joined the research group at Bell Laboratories in New Jersey that was owned by Western Electric and American Telephone and Telegraph Company. The labs carried out research in communication for the Bell Telephone System, originally founded by Alexander Graham Bell in the 1870s. It was then the world's largest industrial research laboratory.

Bell researchers had proposed that semiconductors may offer an alternative to the vacuum tubes or valves then used throughout the telephone system. Shockley conceived several designs based on copper oxide semiconductor materials but attempts to create a prototype were unsuccessful. When World War II broke out, Shockley became involved in radar research as a director at Columbia University's US Navy Anti-Submarine Warfare Operations Group in 1942. He worked to improve the accuracy of Allied attacks on German U-boats and for his contributions was awarded the Navy Medal for Merit.

Shockley returned from his wartime work to start a solid-state physics group at Bell Labs to pursue the research on semiconductor replacements for the bulky, unreliable and power-hungry glass vacuum tubes with electromechanical switches. He conceived a field-effect amplifier and switch, based on germanium and silicon technology, but it failed to work as intended. A year later, theoretical physicist John Bardeen suggested that electrons on the semiconductor surface were blocking penetration of electric fields into the material, negating the desired effects. Working with experimental physicist Walter Brattain, they began investigating the behaviour of these 'surface states'. The key to the development of a new device was their understanding of the process of the electron mobility in the junction of two semiconductors. When p-type and n-type materials are placed in contact with each other, the junction behaves very differently than either type of material alone. Specifically, current will flow readily in one direction but not in the other, creating a one-way device called a *diode*. It was realized that if there was a way to control the flow of the electrons across this diode, an amplifier could be built.

The team started work on building such a device, and tantalizing hints of amplification appeared – sometimes the system worked but then unexpectedly stopped. They found that injecting a very small current in the right place on a single crystal was essential. The so-called emitter and collector leads would both be placed very close together on the top, with the control lead placed on the base of the crystal. When a current was applied to the base lead, the electrons (or holes) would be pushed out across the block of semiconductor and collect on the far surface. If the emitter and collector were very close together on either side of the so-called depletion region, this seemed to allow enough electrons (or holes) between them to allow conduction to start. Finally, they began to get some evidence of power amplification when, acting on a suggestion by Shockley, a voltage was put on a droplet of glycol borate electrolyte placed across a junction.

After numerous attempts, they finally produced the first successful semiconductor amplifier. Bardeen and Brattain had applied two closely spaced gold contacts held in place by a triangular plastic wedge on a crystal of high-purity germanium and then the foil was sliced with a razor at the tip of the triangle. When the plastic was pushed down onto the surface of the crystal and voltage applied to its base contacts, the voltage on one contact modulated the current flowing through the other, amplifying the current. It started to flow by 100 times from one contact to the other as the voltage pushed the electrons away from the base towards the other side near the input signal.

It was on 23rd December 1947 that this new device was first demonstrated to senior managers at Bell Labs. The name chosen for the invention was a *transistor*, being a combination of 'transconductance' and 'varistor'. Their *pnp* 'point-contact' germanium transistor operated as a speech amplifier with a power gain of 18. Despite its delicate mechanical construction, many thousands of units of the early 'Type A' transistor were produced and housed in a small metal package, although the point-contact transistor would prove to be fragile and was difficult to manufacture in high volume with sufficient reliability. It angered Shockley that his name was not on the patent applications, so he secretly continued his own work to build another sort of transistor based on junctions instead of point contacts as he thought his design would be more commercially viable. Shockley was also dissatisfied with certain parts of the explanation for how the point-contact transistor operated so he began an intense theoretical study. By early 1948 he had conceived a distinctly different transistor. He said that positively charged holes could also diffuse through the bulk germanium material and not only trickle along a surface layer – this phenomenon was crucial to operation of his junction transistor. He had invented an entirely new, considerably more robust type of transistor with a layer or 'sandwich' structure. It was made up of a three-layer sandwich of n-type and p-type semiconductors separated by p-n junctions. This structure went on to be used for most transistors well into the 1960s.

Inventors of the transistor at Bell Laboratories (1948)

The next problem was obtaining sufficiently pure and uniform semiconductor materials. This required a critical innovation in chemical technology. Chemist Gordon Teal of Bell Labs had known that large single crystals of germanium and silicon would be required so he suspended a small 'seed' crystal of germanium in a crucible of molten germanium and slowly withdrew it, forming a long, narrow, single crystal. Shockley later called this achievement *'the most important scientific development in the semiconductor field in the early days'*.

It took two more years before scientists and engineers developed processes that allowed Shockley's junction transistor to be manufactured in production quantities. Bell Lab chemists fabricated pn-junctions by dropping tiny pellets of impurities into the molten germanium during the crystal-growing process to form the pnp-structures with a thin base layer. These bi-polar junction transistors soon surpassed the best performance of the point-contact transistors. They consist of three layers of semiconductor called the emitter (E), base (B) and collector (C). The layers are arranged in a sandwich-like structure with two p-type layers surrounding an n-type layer, hence 'pnp'. He had succeeded in making electronic devices much less costly to manufacture. Bell Labs announced Shockley's invention in July 1951. He also published his major treatise *Electrons and Holes in Semiconductors*. It became the reference work for everyone developing new semiconductor transistors, which were going to replace bulky, unreliable vacuum tubes (or valves) in radios, televisions, computers and in many other electronic devices.

The publicity around the coming of the 'transistor age' gave William Shockley a high profile. In 1951, aged 41, Shockley became one of the youngest scientists ever elected to the US National Academy of Sciences and was the recipient of many other awards and honours. In 1956 Shockley with his co-workers John Bardeen and Walter Brattain were jointly awarded the Nobel Prize in Physics for their research on semiconductors and their discovery of the transistor.

In 1952 the transistor was used for the first time in a commercial product, the *Sonotone 1010* hearing aid. Two years later, the first transistor radio, the pocket size *Regency TR1*, was manufactured in Indiana as a joint venture with Texas Instruments, which supplied four germanium junction transistors. Over 100,000 radios were sold, making the word 'transistor' common place. Transistors were soon vital to computer design as they allowed more cost-effective devices to be built that performed faster, were more reliable, used less power and occupied much less space

.

Silicon Valley

To advance his work on semiconductors, Shockley approached a successful businessman for funding. Arnold Beckman in Los Angeles was a Caltech graduate who had invented the electronic pH meter. Shockley moved from New Jersey to California also with the support of Professor Frederick Terman (1900–1982), dean of engineering of Stanford University, to live closer to his elderly mother in nearby Palo Alto. Terman had studied chemistry and took a master's degree in electrical engineering at Stanford University. During World War II, he managed 850 staff at the Radio Research Laboratory in Harvard University, working on countermeasures to significantly reduce the effectiveness of radar-directed anti-aircraft fire. After the war he returned to Stanford and created a microwave research laboratory. In 1951 he launched the Stanford Research Park, whereby the university leased portions of its land to high-tech firms.

Terman greatly expanded the science, statistics and engineering departments, winning research grants from the US Department of Defence. The grants and patent payments generated from research helped Stanford drive the growth of new technology companies in Silicon Valley. Former students at Stanford William Hewlett and David Packard, who had founded Hewlett-Packard in a garage in Palo Alto in 1939, were among the first tenants attracted to the new

Stanford Industrial Park with companies such as Eastman Kodak, General Electric and the Lockheed Corporation.

In 1956, at nearby Mountain View, in Santa Clara County, Shockley founded Shockley Semiconductor Laboratory in a prefabricated former apricot packing shed. As none of his former colleagues at Bell wanted to leave the east coast, he assembled a new team of young chemists, physicists and engineers. Shockley was convinced that the natural properties of silicon meant it would eventually replace germanium as the primary material for transistor construction.

In 1954, although Texas Instruments had started production of silicon transistors, Shockley thought he could still do better. He recruited a creative team including Gordon Moore and Robert Noyce (later Intel's founders), Jean Hoerni (later inventor of a key manufacturing process) and Eugene Kleiner (later a leading venture capitalist). The group began preparing the facility and designing specialized equipment for fabricating silicon devices. They engineered a new type of system that could produce single-crystal silicon ingots (or boules). Shockley taught them how to grow pure crystals and diffuse p-n junctions, leading to a pool of technicians skilled in the science and practise of working with silicon. This was the birthplace of 'Silicon Valley'. Gordon Moore later said: *'Shockley put the silicon in Silicon Valley.'* In subsequent decades, the establishment of other research laboratories, such as IBM in San Jose and the Xerox PARC in Palo Alto, helped to reaffirm Silicon Valley as a leading technological centre, which benefited from its proximity to the Universities of Stanford and Berkeley, to make important advances in hardware and information technology.

Shockley went on to develop a four-layer device known as the '*Shockley diode*' a precursor to the integrated circuit that contained multiple components. However, because of Shockley's sometimes paranoid behaviour, his secrecy and his confrontational management style, eight of his young scientists left him to found the legendary company *Fairchild Semiconductor* in 1957. Shockley later went to Stanford University in 1963 as a professor of engineering and applied science until his retirement.

Integrated circuits

The semiconductor industry developed quickly and within just three years Fairchild's annual sales reached over $20 million; but this was only the beginning of its accomplishments. As the company grew, its employees launched further spin-off businesses. Fairchild Semiconductor became the first major computer 'chip' manufacturer. Its success was based on the revolutionary advances in production techniques for silicon transistors made by its founders. It was a British engineer, Geoffrey Dummer (1909–2002), who had first put forward the concept of reducing transistor components into one solid block or 'integrated circuit' (IC) in 1952, but in 1959 Robert Noyce of Fairchild and Jack Kilby of Texas Instruments independently created the first silicon-based integrated circuit. They had devised a way for manufacturing all the circuit components in a single piece of semiconductor material.

In 1959 Jean Hoerni invented the 'planar' process in which successive layers of material are deposited and patterned by photolithography and etching to create multilayer structures on the surface of the silicon. Silicon dioxide is an ideal electrical insulator, which can be created on the surface of the silicon by thermal oxidation in a furnace or, at later stages in the process, deposited from a chemical vapour. Similarly, metal layers can be deposited in patterns to create complex interconnected circuits. This planar process was licensed by every major semiconductor manufacturer at the time and remains the fundamental approach for making semiconductors. In 1961 Noyce used the process to create the first integrated circuit. The main breakthrough was its ability to deliver miniaturization. The most important application of semiconductors was as logic gates in computers. Fairchild Semiconductor introduced the first practical integrated circuit – a logic gate device about 2 mm in size and sold under the name *Micrologic*.

This first integrated circuit device contained just four transistors; today microchips hold billions of transistors, but they still rely on Hoerni's breakthrough. Perhaps it is the most important innovation in the history of the semiconductor industry. By the mid-1960s, Fairchild, with its new integrated circuits, was generating $90 million in sales. In 1965 Intel founder Gordon Moore remarked that '*microchips will double in power and halve in price roughly every 18 months*'. Moore's 'law' encouraged rapid advancements that enabled more transistors to be put on each microchip at a rapidly declining cost, which drove intense competition between all the innovative computer companies.

In 1964 the Sharp Corporation in Japan produced the world's first all transistor electronic calculator, then a Sharp chemical engineer, Tomio Wada, created the world's first liquid crystal display (LCD) for a pocket calculator in 1973. Optoelectronic materials such as LCD and LEDs for displays and lighting are a special class of semiconductors that either convert electrical energy into light or absorb light and convert it into electrical energy. Sharp's LCD was the ancestor of modern flat-screen displays for TVs, smartphones and many electronic devices. In 1967 Texas Instruments built the first electronic mini-computer using integrated circuits. Advances in integration have kept progressing; for example, they were essential for the successful Apollo moon landing. In 1971 Intel engineers placed 2,300 transistors onto the company's first computer microprocessor chip, the Intel 4004.

Silicon is the second most abundant element on earth and occurs as silicates and silica. Silica is the main component of sand or quartz. The chemical technology for the industrial fabrication of semiconductors is extremely complex, involving high-purity materials, sophisticated equipment and hundreds of steps. It is the most complex industrial mass production process ever devised. It begins with the growth of a large single crystal or ingot of silicon, sliced into wafers on which silicon chips are built and arranged in a grid formation. A polished wafer is typically 300 mm in diameter and only 0.775 mm thick. The doping process whereby special impurities (for example phosphorus) are added onto the silicon to give it the desired capabilities is followed by the creation of numerous components on the chips, which could number into millions. Chips are fabricated using photolithography and etching processes. Contamination from dirt and dust can be detrimental to the process. It is for this reason that the whole process is undertaken in spotless and air-filtered facilities called 'clean rooms'.

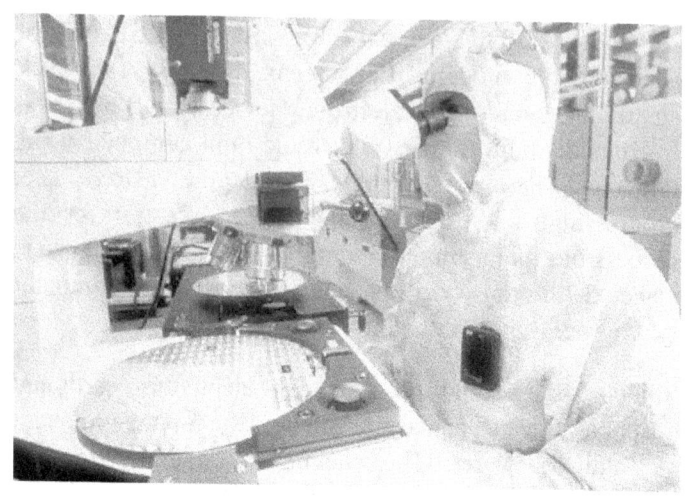

Testing silicon wafers in a semiconductor plant clean room

By 1993 Intel's first *Pentium* microprocessors contained over 3 million transistors – a remarkable technological achievement. The delivery of faster processing speeds coupled with the demand for increased miniaturization have been the greatest challenges facing the modern semiconductor industry. In 2020 Apple Inc. announced that it had produced the new *Mac M1* processor chips for its computers containing 16 billion transistors. The demand to produce chips with ever greater numbers of transistors and complexity, has required photolithography processes printing patterns using very short wavelengths of light such as EUV (extreme ultraviolet) well beyond the visible light spectrum. This light is produced by using a high-powered laser that impinges upon droplets of tin to create plasma emitting light at a wavelength of 13nm. EUV silicon-chip fabrication machines are extremely costly to build, at over $100 million.

Critical technology

The semiconductor industry, which makes vital components for the technologies everyone depends on has become critical to the smooth functioning of the global economy. In many ways, the modern world is built of silicon semiconductors. Shockley's and then Fairchild's successes continued to fuel the growth of many companies world-wide. The electronic semiconductors created first in the late 1940s have become the core building blocks of thousands of new innovations, profitable businesses and led to a new industrial revolution. They spawned hundreds of ventures that established Silicon Valley as a world centre of entrepreneurial activity and technological leadership. Much of the expertise spread to Asia and Europe but perhaps over 70% of today's high-technology companies in the valley on US stock exchanges (such as Google, Alphabet, Meta, Pixar, YouTube, Amazon, Microsoft and Cisco Systems) have been derived from the original pioneering entrepreneurs.

As the impact of digital technology on lives and businesses has accelerated, semiconductor sales have grown to over $600 billion globally. About 70% of future growth in the medium term is predicted to be driven by three industries: automotive, computation, and data storage – to support artificial intelligence (AI), cloud computing and wireless 5G smartphones. Today, East Asia companies produce 90% of all memory chips and 75% of the logic chips used globally with Taiwan making more than 90% of the most advanced chips.

It was said that World War II was decided by those nations that had more steel and aluminium, but in future the rivalry between superpowers will depend on dominance in computing power and having control of the production of the cutting-edge semiconductors used for machine learning, communications and military hardware.

F: Fuel cells

Apollo spacecraft orbiting the Moon

Electrochemistry reaches the moon

Fuel cells are a power source, which like batteries use electrochemical reactions to convert energy stored in chemical bonds into electricity. They differ from batteries in that the reactants are continuously supplied, rather than being stored internally, and hence can operate without recharging. Fuel cells use gases, typically hydrogen, and oxygen to produce the electricity. This can then power electric motors, for example in vehicles, or store energy in batteries for later use. In broad terms warm air and clean water are the only by-products of a hydrogen fuel cell, making them 'zero-carbon' emitters and a sustainable power source. Fuel cells were first created in the 1930s then had a critical role in the space race on board spacecraft going to the moon from the 1960s. They are now part of the

21st-century revolution in new energy sources to reduce dependence on fossil fuels and to cut emissions of carbon dioxide and other greenhouse gases.

A simple fuel cell is composed of two electrodes, a cathode and an anode, separated by a porous electrolyte membrane. Hydrogen gas enters the fuel cell through the anode where the atoms of hydrogen, in combination with a catalyst, split into protons and electrons. At the same time, oxygen from the air enters the fuel cell through the cathode. The protons pass through the membrane while from the other side the electrons flow out of the cells, creating an electrical current. At the cathode, the oxygen and protons then combine to produce water. Unlike thermal energy generators, fuel cells do not have efficiency limits based on size – no matter how small the unit. It is their modular design and efficiency that makes them suitable for a wide range of applications.

The principles of fuel cell operation were first demonstrated in the 19th century, but was not revisited for almost 100 years until it was revived by a British engineer. In 1932 Tom Bacon developed the world's first practical hydrogen fuel cell. His alkaline fuel cell (AFC), also known as the 'Bacon' fuel cell, would later form the power source used in the Apollo moon landing mission.

Considering the time when global growth would deplete oil and gas reserves, Bacon wrote in 1972: *'The only real synthetic substitute is likely to be hydrogen.'* The prediction that hydrogen would become a viable alternative to oil remains valid, with some forecasts predicting that by 2050 hydrogen fuel cells will power around a third of the world's vehicles. Bacon spent his whole working life on the development of the hydrogen fuel cell.

Pioneer

Francis 'Tom' Bacon (1904–1992) studied mechanical engineering at Cambridge University, which helped him later as he went into chemical engineering and reactor design. After graduation he became an apprentice with Parsons, the Newcastle engineering firm that manufactured steam turbines. A few years into his job, he found himself intrigued by an idea for electrolysing water with electricity and using the resulting hydrogen and oxygen to power a vehicle.

The concept of the fuel cell had been first demonstrated by Sir William Robert Grove in 1839. Bacon thought his engineering experience of working at high temperature and pressure would be useful, although his company did not think fuel cell research was relevant to its business. Unperturbed, Bacon proceeded to experiment at home using a platinum gauze as a catalyst in sulfuric acid then activating nickel electrodes separated by a pure asbestos cloth with an electrolyte of aqueous potassium hydroxide, operating at high pressures and at over 100 °C. He initially devised a reversible cell that would first split the water into hydrogen and oxygen and then (in the same cell) recombine the gases to produce electricity, though he soon modified the design to a two-cell system where the electricity was successfully generated in a separate chamber to the hydrogen production.

However, while these early experiments worked, the corrosive chemicals, high pressures and temperatures involved made them unsuitable for being carried out on a table at home, so Bacon secretly moved his equipment to his office at work. When his employer found out, he was told to stop working on the fuel cell or return to academia. By 1946 he had gone back to the University of Cambridge's Department of Chemical Engineering.

Developing a usable fuel cell was an arduous task, not least because when operating at temperatures of 200 °C and pressures of up to 40 bar (4 MPa), sealing gaskets were prone to deteriorate and the asbestos diaphragm frequently failed. In later years, Pratt & Whitney engineers joked that: *'if a screwdriver were to be dropped into a vessel containing the hot electrolyte for a Bacon cell, it would dissolve before it hit the bottom!'* But even with such extreme chemistry, Bacon's target of producing a usable potential was proving to be elusive.

A key problem was establishing a stable interface between the hydrogen or oxygen and the electrolyte. At the suggestion of Eric Rideal an expert on colloids, Bacon developed a bi-porous membrane with relatively large pores on the gas side and much finer ones for the electrolyte, which solved the interface problem and eliminated the fragile asbestos diaphragm. The performance of the cells improved significantly but they had a short lifespan, as the electrodes corroded and the nickel-plated cells buckled and collapsed. The problems with corrosion of the oxygen electrode were overcome by soaking the nickel electrodes in lithium hydroxide solution followed by drying and heating. The hydrogen catalyst poisoning was solved with the use of PTFE rings. Bacon was able to present a working six-cell fuel cell (producing 0.8 V per cell at 230 mA/cm^2) at an exhibition in London. Unfortunately, there was little commercial interest in his early fuel cell, so he sought funding from government to produce a larger cell.

His fuel cells were to find a critical role in the space race on board US spacecraft bound for the moon in the 1960s. NASA had given Pratt & Whitney the project to develop a source of electricity for the Apollo mission. They licensed Bacon's patented technology because fuel cells were ideal for spacecraft, given hydrogen and oxygen were already used for rocket propulsion and life support. Unlike conventional heat engines, whose efficiency falls with decreasing load, that of the fuel cell rises, so it was possible by restricting electrical loading to operate at 75% efficiency. Additionally, the by-product water could be used for drinking and humidifying the atmosphere of the space capsule.

Having few financial constraints, Pratt & Whitney assembled a team of 1,000 engineers and spent the huge sum of $100 million on development (over $800 million today). The Apollo 11 mission was a great success and astronaut Neil Armstrong became the first person to walk on the moon. Bacon received a personal letter of thanks from company engineers: *'The three power plants performed flawlessly and provided about 400 kWh of electrical energy during the mission. Your satisfaction must be very great that your pioneering efforts have made this possible.'*

Bacon met Armstrong and his fellow astronauts when they visited Britain. US President Richard Nixon put his arm around Bacon and said: *'Without you, Tom, we could not have gotten to the moon.'* He was a modest man and he tended to credit the engineers at Pratt & Whitney. Yet without his perseverance there would not have been a highly efficient fuel cell in the spacecraft. As a result of the success of his design, the fuel cells were used for all the subsequent manned spaceflights and space shuttles. A chemical engineering colleague at Cambridge remembered Bacon's enthusiasm for a hydrogen-based economy saying: *'He was a man 50 years ahead of his time. He had old-fashioned charm, was very approachable and wore an old school tie.'* He recalled his determination: *'He was in the lab every day of the year. I do not think I have met anyone more single-minded.'*

Although Tom Bacon died in 1992 aged 88, two decades before vehicles powered by fuel cells could be a reality, his AFC became one of the most developed fuel cell technologies. Bacon was an ardent supporter of the idea of using fuel cells for road transport, as he had realized that their high efficiency and low pollution made them a highly desirable energy source.

Bacon had keenly supported the use of hydrogen as an alternative fuel, recognizing its special safety and infrastructure needs. Hydrogen like electricity is a zero-carbon energy carrier (no emissions at point of use). It can be used to power high-efficiency fuel cells, to provide energy storage, as a supplement or replacement for natural gas and as a vehicle fuel. Currently, hydrogen gas needs to be produced from chemical technology – most commonly from hydrocarbon feedstocks such as methane (natural gas) that produce carbon dioxide emissions, which must be mitigated by using carbon capture and storage. The 'green' way of producing hydrogen is by electrolysis using renewable electricity to split water molecules. The key challenges for electrolysis are reducing its capital cost, securing sufficient low-carbon electricity and improving the conversion efficiency.

A few novel hydrogen production methods are also at the early stage of development, such as bio-hydrogen from algae, photocatalytic hydrogen production from sunlight and water and the high-temperature sulfur-iodine cycle, which splits water directly into its elements. Hydrogen can also be used as a storage medium for electricity since it can be produced, stored and then used to later generate electricity in fuel cells or gas turbines.

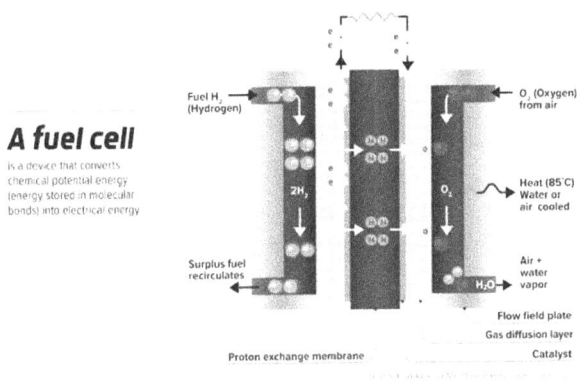

A fuel cell

is a device that converts chemical potential energy (energy stored in molecular bonds) into electrical energy

Fuel cell construction

Types of fuel cells

There are several types of fuel cells, each defined by the type of electrolyte they use as well as their operating temperature. The choice of which type to use is normally based on the fuel cell's targeted application.

AFCs use a liquid potassium hydroxide electrolyte. The catalyst can be easily poisoned, so the oxygen and hydrogen used in AFCs need to be purified, which adds to its cost. Phosphoric acid fuel cells (PAFC) utilize phosphoric acid as the electrolyte, which requires a very high temperature to start the reaction and uses platinum as a catalyst. Although costly, PAFCs are more tolerant to impurities and are very efficient at producing heat and electricity, making them ideal for power plants. Proton exchange membrane fuel cells (PEMFC) employ a platinum-coated solid polymer at low temperatures and require only oxygen and hydrogen to generate electricity, so they are lighter and have high-power density. Finally, direct methanol fuel cells (DMFC) use methanol rather than pure hydrogen, which is a safer and much easier fuel to store and transport.

Hydrogen power

The global economy faces a significant transition to alternative energy technologies from fossil fuels to reduce greenhouse gas emissions and air pollution (see Z). The future of hydrogen fuel cells looks promising as a viable option for power generation, especially in applications such as shipping, heavy goods or public transport. The very reliable individual fuel cells can be combined to generate power in large-scale installations, can act as uninterruptible power supplies (UPS) in protecting data centres and other critical infrastructure from supply failures and can supply power in locations far removed from grid access. Fuel cells have already been used for restoring remote power stations to operation without relying on an external power transmission grid. Of course, hydrogen currently remains relativity costly, being highly flammable and difficult to transport and store safely on a large scale.

When hydrogen fuel is produced by electrolysis powered from renewables the fuel cell emissions consist of only water. These cells possess many advantages being clean sources of power, not requiring a recharge period like batteries and operate with very little noise or maintenance as they have no moving parts.

G: Gunpowder and explosives

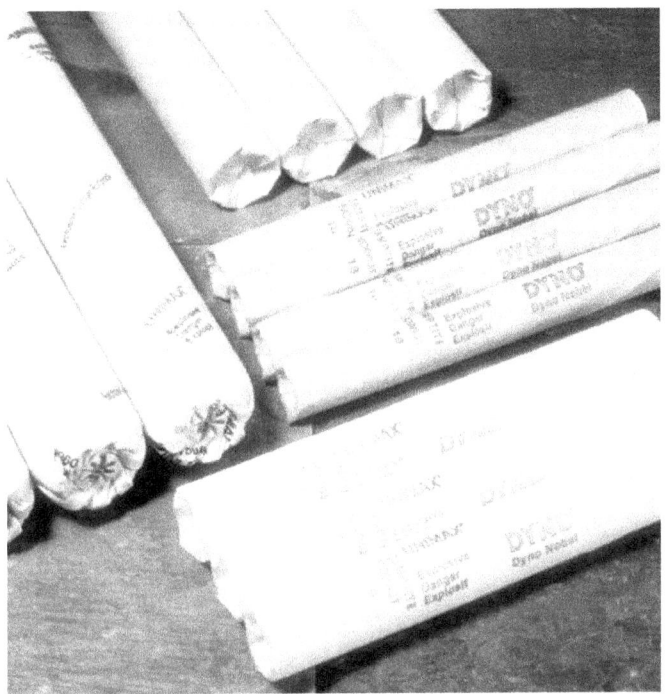

Sticks of dynamite from Dyno Nobel Inc.

Moving mountains

Gunpowder, also known as 'black powder', was the earliest chemical explosive and is considered one of the first applications of chemical technology. It made possible more effective and powerful military weapons. Explosives are substances that produce a massive volume of rapidly expanding gas in an extremely brief period. In the 19th century a new explosive – *dynamite* – was created that revolutionized construction activities by reducing the costs of blasting rock to make way for highways, railways and bridges. It also allowed the mining and quarrying of essential raw materials on a much larger scale.

Gunpowder is made from a mixture of sulfur, carbon and saltpetre (potassium nitrate), the sulfur and carbon from charcoal acting as fuels and the saltpetre an oxidizer. Gunpowder has been widely used as a propellant in firearms, artillery, rocketry and pyrotechnics but also as a blasting agent for other explosives in quarrying, mining and road building. It is classified as a low (or deflagrating) explosive because of its relatively slow decomposition rate and consequently low detonation pressure, burning at *subsonic* speeds. In contrast, high explosives such as dynamite and TNT [2,4,6-trinitrotoluene] are characterized by extremely rapid decomposition and the development of high pressure, producing a *supersonic* shockwave. Some explosives require a detonator or a supplementary booster. Gunpowder is used as a propellant in firearms as its ignition behind a bullet generates pressure to force the shot from the muzzle at high speed, but not enough force to rupture the gun barrel. It was widely used in fused artillery shells until the first detonating or high explosives were developed in the second half of the 19th century. Gunpowder is less suitable for shattering rock or fortifications with its low explosive power. The new chemical technology making powerful explosives such as dynamite became essential to the mining and engineering industries.

Dynamite was the invention of Alfred Nobel, a Swedish chemist, engineer and armaments manufacturer, famous for endowing his enormous fortune to institute the Nobel Prizes.

History

The earliest surviving written description of making an explosive – as an accidental by-product of Chinese alchemy to develop an 'elixir of immortality' – dates from the 9th century. The book *Classified Essentials of the Mysterious Tao of the True Origin of Things* warns: *'Some have heated together sulphur, realgar (arsenic sulphide) and saltpetre (potassium nitrate) with honey; smoke and flames result, so that their hands and faces have been burnt, and even the whole house where they were working burned down.'*

The Chinese began to use crude gunpowder for incendiary projectiles or fire lances in warfare. The *Wujing Zongyao*, a Chinese military compendium written around 1040, specified one of the first explosive recipes for incendiary bombs thrown by siege engines as containing 50% saltpetre, 25% sulfur, 6% charcoal and 19% other materials. One document describes a bomb launched from a catapult containing small, powder-filled spiked iron fire-balls designed to stick to targets and set them alight. A fire lance was also used, known as *Tu Huo Qiang*, it combined a long spear with a simple firework-like charge at the end underneath the spear head. The weapon, considered a key stage in the development of firearms, would shoot out the projectile along with flame in close combat. These lances using gunpowder for warfare paved the way for the development of other forms of weaponry – the gun, the rocket and cannon. The Chinese probably began to use rockets in war in the middle of the 13th century. Saltpetre explosives developed into an early form of black powder in China, with the technology spreading west from China to the Middle East and then onto Europe.

Chinese fire-lances using gunpowder - predecessors to guns

Some scholars believe that while the Chinese developed explosives for use in fireworks, it was Arabs who first employed gunpowder in weapons. In Britain, King Edward III first used it in an early form of cannon about 1327. Larger artillery pieces, popularly known as bombards, were in use across Europe by the 15th century. By the end of the 18th century gunpowder was composed typically of 75% saltpetre, 15% charcoal and 10% sulfur. In 1845 Christian Schönbein, a German-Swiss chemist, invented *guncotton* by dipping cotton in a mixture of concentrated nitric and sulfuric acids and then washing with water. This produced a propellant for military weapons as a substitute for gunpowder, but it was too rapid and violent. In 1846 the Italian chemist Ascanio Sobrero prepared *blasting oil*, a potentially dangerous mixture of glycerine, concentrated nitric and sulfuric acids known as *nitroglycerine*.

During the late 19th century, the use of gunpowder in warfare declined. The invention of *smokeless* powder combatted gunpowder's disadvantages in warfare – being sensitive to moisture and producing quantities of dark smoke that identified the solider firing. French chemist Paul Vieille made the first smokeless powder in 1884. He partially dissolved nitrocellulose in a mixture of ether and alcohol until it became a gelatinous mass, which he rolled into sheets and then cut into flakes. When the solvent evaporated, it left a hard, dense material that gave satisfactory results in all types of armaments.

Pioneer

The leading explosives pioneer was Alfred Nobel (1833–1896), who was born in Stockholm, Sweden. His father, Immanuel, was an engineer, industrialist and inventor who was involved in a business constructing railroads and bridges in Russia. In 1842 the family moved to St Petersburg, where the young Alfred Nobel was educated by tutors in chemistry and physics. In 1850 Nobel was sent to Paris to further his education in chemical technology. There he studied under the famous chemist Théophile-Jules Pelouze at the University of Torino and met Sobrero, the inventor of nitroglycerine. Nobel helped his father in the family factory, which during the Crimean War (1853–1856) produced munitions for the Russian military. He later developed and manufactured naval mines that successfully protected the ports from invasion by the British Navy during the war.

After the war, the Nobels reorganized the firm and entered the oil industry, in which they also prospered. Alfred's first major invention was a blasting cap or detonator, a wooden plug filled with black gunpowder, which could be detonated by lighting a fuse. They established a factory in Heleneborg,

Sweden, to capitalize on their innovation. They were also the first to produce nitroglycerine on an industrial scale. It was the speed of decomposition that made the nitroglycerine a powerful explosive. In the reaction, 4 moles of nitroglycerine rapidly produced 35 moles of gases. The shock from the initial decomposition reaction created a supersonic pressure wave that propagated through it, causing near-instantaneous pressure-induced decomposition of the surrounding fuel in an explosive blast. While nitroglycerine was many times more powerful than gunpowder, it was so volatile that it could not be safely used in blasting as it was extremely unstable and sensitive to shock. It was prohibited in many countries, owing to its extremely dangerous nature and the frequency of accidents during its use. After several serious explosions, including one in 1864 that killed his brother Emil, Alfred Nobel was convinced that a method to make nitroglycerine safer to use had to be found.

At his laboratory near Stockholm, Nobel found a way to improve the powerful explosive. In 1867 he made one of his most important discoveries: that by mixing three parts of the oily nitroglycerine fluid with one part of *kieselguhr*, a diatomaceous earth containing fossilized remains of tiny sea organisms, the mixture could be turned into a paste he named *dynamite*. The mixture was stable enough to be handled without the risk of accidental detonation until ignited with a mercury fulminate cap. He also added a small amount of sodium carbonate to neutralize trace quantities of acids that formed during storage. This material could be kneaded and shaped into rods suitable for insertion into the drilling holes for mining. Nobel patented his discovery in 1867. Dynamite is a blasting explosive not a propellant. The introduction of dynamite greatly improved operational safety and reduced the cost of rock blasting, building canals and demolition work.

As a result of Nobel's discovery there was an immediate expansion in blasting activities worldwide. After establishing factories in Germany and France in 1871, Nobel went to the UK to build a successful British company (British Dynamite), which became the largest explosives works in the world. He chose a huge isolated site among sand dunes at Ardeer on the west coast of Scotland.

A contemporary report described the initial process there:

> *The highly explosive nitro-glycerine was prepared in buildings located between 'hills' (isolated earth mounds). In each of the nitrating buildings are two huge cylindrical lead tanks with dome-shaped tops, five feet in diameter by six feet in depth. Beside the vat is seated the operative on a one-legged stool (in case he fell asleep!), with his eye glued to a long thermometer extending into the interior of the cylinder. The vat, cooled by a cold-water jacket and coils, is charged with sulphuric acid mixed with nitric acid. The glycerine is carefully admitted in the form of a fine spray, the jet being maintained by compressed air. The full charge within the cylinder of the nitric and sulphuric acids must be supplied with 900 pounds (400 kg) of glycerine, generating intense heat being attended by the ever-present danger of spontaneous explosion.... The nitro-glycerine was then made into dynamite by combining it with kieselguhr earth in a mixing box.*

Powerful new explosives

In 1875 Nobel developed an even more versatile and powerful explosive for blasting. *Gelignite* consisted of nitrocellulose dissolved in either nitroglycerine or nitroglycol, mixed with wood pulp and saltpetre. It was more stable than dynamite and its composition made it more mouldable and safer to handle. Next, he invented a smokeless blasting powder called *Ballistite*, made from nitroglycerine, nitrocellulose and camphor, which gave a propellant with a powerful but more controlled rate of burn that did not threaten to damage weapons. A British version of gelignite for the military, *cordite*, was developed by Sir Frederick Abel, a renowned government chemist and Professor James Dewar from the University of Cambridge, as the British found fault with Nobel's patent.

These new explosives revolutionized mining and civil construction by reducing the cost of blasting rock to build highways, railroads and bridges. At the same time, they also contributed to increasing the destructive force of military weaponry. Nobel's inventions and innovations brought him great wealth, but criticism of his manufacture of armaments made him reflect in later in life on more peaceful endeavours. In his will, he set aside the bulk of his estate to establish the Nobel Prizes. The prizes were to be awarded annually to those who significantly added to the progress of humankind in science, medicine, literature and peace. Alfred Nobel's full contribution to science was not confined to munitions or explosives. His work led to the manufacture of artificial silk, rubber and semiprecious stones while researching in optics, electrochemistry and biology. In addition to being a creative inventor, he had commercial flair and laid the foundations for the multinational company. He died in 1896 at his villa in San Remo, Italy.

The organic chemicals industry had begun with synthetic dye making based on materials derived from coal tars. The first coal tar explosive was picric acid, called *Lyddite*. It was originally used as a yellow dye and made by nitration of phenol. When another coal tar chemical, toluene, was nitrated to produce TNT it became one of the most widely used explosives. Its explosive properties were first discovered by the German chemist Carl Häussermann in 1891. It was safe to handle, insensitive to shock and could be safely poured into shell cases, but it exploded with great force as the nitro-groups in the molecule rapidly turned into nitrogen gas. To detonate, TNT must be triggered by a pressure wave from a starter explosive, called an 'explosive booster', which itself is initiated by a detonator, so it is not prone to accidental detonation, unlike the sensitive nitroglycerine. The German armed forces used TNT for making artillery shells that would explode after they had penetrated the armour of ships, whereas the older Lyddite-filled shells tended to explode upon striking armour outside the ship. The US Navy filled its armour-piercing shells with ammonium picrate known as *Dunnite* or *Explosive D*.

Many developments in explosives were to follow in the years after World War I. Britain's explosives manufacturers were finally combined under Nobel's company and later became part of the major chemical company ICI.

Explosive growth

One of the most significant changes in the explosives industry since the invention of dynamite was the development of *ammonium nitrate fuel oil* mixtures (ANFO) and ammonium nitrate water gels. These were patented in 1955 and are typically composed of 94% pellets of ammonium nitrate, which acts as the oxidizing agent and absorbent for the fuel, and 6% fuel oil. ANFO is detonator insensitive, so it must be initiated by a primer. It comes in either bulk form, which is mixed and pumped directly into boreholes, or pre-mixed bags. It has found wide use in quarrying, coal and metal mining, and civil construction, where its low cost and ease of use may outweigh the benefits of other high explosives.

After gunpowder, chemical technology developed advances in explosives that helped to facilitate building the modern world's vast infrastructure and allowed easier access to essential minerals, ores and raw materials. Controlled explosions are employed to create tunnels, remove overburden and facilitate the extraction of resources. High explosives like TNT and RDX are commonly used in military applications, while low explosives such as gunpowder and smokeless powder still find uses in propellants and fireworks. The production, storage and handling of explosives require strict safety measures to prevent accidents and unauthorized use.

The global explosives market, valued at $20 billion, is experiencing steady growth driven by industrialization, infrastructure development, mineral exploitation and military warfare or defence modernization.

H: Hydrocarbon processing

Pulau Bukom refinery and petrochemicals complex Singapore

Making oil useful

The processing of hydrocarbons has provided massive benefits for modern living. The light hydrocarbons derived from crude oil are firstly valuable energy sources due to their high energy density, transportability and relative abundance. Over 80% of the hydrocarbons from oil are converted into energy-rich fuels such as petrol, kerosene and aviation fuel or used for heating. The development of an oil-based or petroleum industry from the late 19th century arose from discovering oil reserves combined with advances in chemical technology for hydrocarbon processing. Among those significant innovations in hydrocarbon processing were both the Houdry fixed-bed catalytic process and fluidized-bed catalytic cracking (FCC).

Apart from fuels, the processing and refining of crude oil and natural gas allowed the production of many petrochemicals that form the basic building blocks for a large variety of essential products. Petrochemicals, such as ethylene, propylene, butadiene and benzene, are used for making plastics such as polyethylene, PVC and polypropylene that are widely used in everything from cars to buildings, food packaging to fabrics and flooring or are used to produce countless other synthetic materials and products.

The hydrocarbon feedstocks of petroleum oil, coal and natural gas are fossil fuels and considered non-renewable energy sources. The known reserves of oil or petroleum are estimated to be over 1,500 billion barrels (oil barrel: 160 litres) so that with global consumption currently around 35 billion barrels per year, all known oil, at this rate, could be consumed by 2060. As fossil fuels are burned, they release carbon dioxide, which as a greenhouse gas is now known to be impacting global climate. As fossil fuel reserve supplies reduce and policies to limit their use are introduced, their importance will decline. Now, finding economic alternatives to petroleum is crucial to global energy security and the focus of many chemical technologists.

Formation

Crude oil or petroleum (Latin: *petra* – rock; *oleum* – oil) is found in vast underground reservoirs where ancient seas were located and consists of a complex mixture of hydrocarbons (mostly alkanes of various chain lengths). Shorter hydrocarbons are gases or low-boiling liquids, while long-chain hydrocarbons are more viscous liquids. Crude oil is usually black or dark brown but varies greatly in appearance, depending on its composition. Those containing fewer metals or sulfur, for instance, tend to be lighter and clearer. Crude oil may also occur in semi-solid form mixed with sand, as in the Athabasca oil sands in Canada, a crude bitumen. The biogenic theory suggests oil and natural gas were formed by compression and heating of ancient remains of prehistoric zooplankton and algae that settled to the bottom of the sea in large quantities under low oxygen (anoxic) conditions millions of years ago. Decaying terrestrial plants and forests, on the other hand, tended to form coal seams. Over geological time the organic matter was buried under heavy layers of sediment, which produced high levels of heat and pressure and caused it to chemically change, first into a waxy material known as kerogen and then with more heat into liquid and gaseous hydrocarbons. As hydrocarbons are lighter than rock or water, they migrate upward through adjacent porous rock layers until they become trapped beneath impermeable rocks in reservoirs, from which the liquid can be extracted by drilling and pumping. A vast amount of the oil escaped to the surface and has been biodegraded by oil-eating bacteria. Oil companies are exploring for the small fraction that was trapped in accessible rock strata.

Discovery of oil

The invention of the first modern steam engine, in the 18th century, heralded Britain's transformation to an industrial economy powered by coal. Coal produced four times as much energy as the same amount of wood and was cheaper to produce and, despite its bulk, easier to distribute. Coal-fired steam pumps increased mine output, while steam-powered locomotives and steamships dramatically reduced the time and cost of transportation. Coal-powered engines in textile mills enabled a breakthrough in productivity.

The dawn of the 20th century was to open a new technological era powered by oil as the new energy source. Oil deposits were known to the Chinese long ago, and the earliest oil wells were drilled using drill bits attached to bamboo poles in the 4th century. The oil was used as a fuel to evaporate brine to produce salt. By the 10th century, extensive bamboo pipelines connected oil wells with salt springs. In Persia, the streets of Baghdad were sometimes covered with tar, derived from petroleum found leaking from nearby oil fields. The distillation of petroleum was first undertaken by Persian chemists for producing fuel for lamps or for military purposes. Explorer Marco Polo had observed oil fields near Baku, Azerbaijan. These techniques spread to Europe by the 12th century. The first patent for extracting oil from oil shale was granted in 1694 to the British entrepreneurs Thomas Hancock and William Portlock, who reported they had *'found a way to extract and make great quantities of pitch, tarr, and oyle out of a sort of stone'*.

The modern hydrocarbon processing industry can be said to have begun when Abraham Gesner (1797–1864), a Canadian geologist, developed a process to refine a liquid fuel from coal, bitumen and oil shale. His discovery, which he named *kerosene*, burned more cleanly and was less expensive than competing products, such as whale oil. In 1850 Gesner created the Kerosene Gaslight Company and began installing lighting in the streets of Canadian towns. By 1854 he had expanded to the US, where he started a company on Long Island in New York. The industry grew as demand for lighting increased, but expansion was restricted until discovery of larger petroleum reserves. A given quantity of oil had a calorific value some 40% more than coal. A commercial oil refinery in Pittsburgh was established by Samuel Kier in 1854 using crude oil from nearby salt wells to produce an illuminating oil. Polish inventor Ignacy Łukasiewicz improved Gesner's method and discovered a means of refining kerosene from the more readily available 'rock' oil, and the first rock oil mine was built in southern Poland in the following year. In 1846 the first oil well in the world was hand-drilled in Russia by Major Alekseev of the Mining Engineer Corps of Baku. Small oil refineries were also opened in Poland and then in Romania. The refinery in Romania covering four hectares had a daily production of over seven tonnes and employed wood-fired iron vessels. It was one of the world's first profitable oil refineries, allowing Bucharest to become the first city in the world illuminated entirely with distilled crude oil.

In Britain, modern petroleum processing began in 1847 when Scottish chemist James Young found a deposit of natural petroleum in a colliery at Alfreton, Derbyshire, from which he distilled a light oil suitable for use as lamp oil and a thicker oil suitable for lubricating machinery. His new oils were successful but supplies of oil from the coal mine soon dried up. Young, noticing that the oil was dripping from the sandstone roof of the coal mine, conjectured that it originated from the action of heat on the coal seam and so thought that it might be produced artificially. He tried many experiments and eventually succeeded, by distilling oil shale on a low heat to produce a petroleum liquid. Young found that by slow distillation he could obtain a few useful liquids, one of which he named 'paraffine oil' because at low temperatures it congealed into a substance resembling paraffin wax. The production of these oils and paraffin formed the basis of his patent of October 1850. The following year, Young founded a substantial shale oil industry at Bathgate, West Lothian, in Scotland and his works became the first truly commercial oil refinery in the world. Oil was extracted from locally mined shale and bituminous coal to manufacture naphtha and lubricating oils and paraffin for fuel use and in solid form.

By 1865 some 120 shale-oil works were operating in Scotland, reaching a peak in 1913 when over 3.2 million tonnes of shale was being processed annually. The industry inevitably declined as imports of cheap crude oil from the Middle East became available. The first oil refinery was built in France around 1857 near an oil sand mine in Alsace, which subsequently became the birthplace of modern global oil field services company Schlumberger.

In 1857 the first mechanically drilled oil well in the world was drilled by the American Merrimac Company in La Brea in south-east Trinidad. The first oil well in North America was constructed by James Miller Williams, an asphalt producer in Oil Springs, Ontario, Canada, in 1858. Williams was drilling for water but was aware that local indigenous people had already used the sticky oil deposits to waterproof their canoes.

The modern oil production industry is said to have truly begun in 1859 when Colonel Edwin Drake drilled an oil well on Oil Creek near Titusville, Pennsylvania, for the Seneca Oil Company. Edwin Drake was born in 1819 at Greenville, New York, spending his early life working the railways in Connecticut for the New York and New Haven Railroad. He took his early retirement in Titusville, Pennsylvania, where Seneca Oil, originally called the Pennsylvania Rock Oil Company, had demonstrated the significant economic value of petroleum. The company hired Drake to investigate the oil seeps on its land. He purchased a steam engine to power a drill and with William Smith, a blacksmith and experienced salt-well driller, made the tools to carry out the drilling. The well was dug on an island in Oil Creek. It took some time for the drill to get through the layers of gravel, and at 5 metres deep the sides of the hole began to collapse. Many workers began to despair, but not Drake, as he devised the idea of a 'drive' pipe. This cast-iron pipe consisted of 3-metre-long joints. The pipe was driven down into the ground and at about 10 metres they struck bedrock. The drilling tools were then lowered through the pipe and a steam-powered drill was used to penetrate the bedrock. The going was slow – a rate of just 1 metre per day. After their initial difficulty, Drake and Smith finally proved successful. Reaching a depth of 21 metres on 27th August 1859, the drill slipped into a crevice 15 cm below the drilled hole. Smith pulled up the tools and headed home. The next day when he went back to the well, he discovered oil floating on the water not far from the derrick. Drake set up a company to extract and market the oil. By end of the year, output was running at the rate of 15 barrels (2,400 litres) per day.

This pioneering work led to the growth of the modern oil industry and set the stage for a new wealth-creating economy. Unfortunately, Drake failed to patent his drilling invention and lost all his savings in oil speculation, dying an impoverished man in 1874. In 1875 crude oil was discovered by David Beaty at his home in Warren, Pennsylvania. This led to the opening of the Bradford oil field, which, by the 1880s, produced 75% of the global oil supply. The legendary businessman John D. Rockefeller, who founded Standard Oil in 1865, was by 1880 controlling 90% of America's refining and distribution. Standard Oil's dominance grew to cover exploration, production, refining and consumer sales. In 1901 a major oil gusher at the Spindletop field in south-eastern Texas sent more than 800,000 barrels of crude into the air before it could be brought under control. The Spindletop 'strike' helped increase the annual oil output in the US from a mere 2,000 barrels (270 tonnes) in 1859 to more than 65 million barrels (8 million tonnes) by 1901.

While Rockefeller was building his empire in the US, the Nobel and Rothschild families were competing for control of production and refining in Russia. In search of a global transportation network to market their kerosene, the Rothschilds commissioned the first oil tankers from British trader Marcus Samuel. He formed Shell Transport and Trading in 1897, which later combined with Royal Dutch Petroleum in the Dutch East Indies to integrate production, pipeline and refining operations. In 1912 Royal Dutch Shell purchased the Rothschilds' Russian oil assets. Earlier in 1908 the discovery of oil in Iran by a British former gold miner, William Knox D'Arcy, in co-operation with a Middle Eastern Shah, led to the founding of Anglo-Persian Oil Company. The British government purchased a majority share of the company in 1914 to ensure sufficient oil for the Royal Navy. It later became British Petroleum (BP).

By the early 20th century, oil emerged as the preferred energy and raw material source due to the rapid growth in use of petrol or gasoline for the internal combustion engine in cars and vehicles. By 1919 petrol demand exceeded that of kerosene as the oil-powered ships, trucks, tanks and military airplanes in World War I had proved its strategic importance. Prior to the 1920s, any natural gas that was produced along with oil was burned (or flared) as a waste by-product. Eventually, natural gas would be a valuable feedstock and fuel for industrial and residential heating and power generation. During the 20th century major crude oil reserves were discovered in Kuwait, Saudi Arabia and Libya, then later in South America, Iraq, Nigeria, Canada, the North Sea and Gulf of Mexico.

Oil processing

Crude oil must be refined and processed before it is used to manufacture products such as petrol, heating oil and plastics. As the demand grew for petrol, chemical engineers developed refining techniques and discovered a host of useful by-products, from which a new petrochemical industry grew. In a refinery, the various components in crude oil were first split up using distillation by applying heat to separate it into fractions, within certain boiling ranges. The quality of these initial fractions was not sufficient to be sold directly as petroleum products without further processing, which often converted unwanted heavy fractions into marketable products.

In 1908 a new method of oil refining that significantly increased petrol or gasoline yields, known as *thermal cracking*, was developed by the chemical engineers William Burton and Robert Humphreys of Standard Oil. They discovered that by applying heat and pressure during distillation, petroleum crude molecules broke down into smaller molecules. The heavier molecules were split or 'cracked' into the lighter petrol molecules that were required by the growing vehicle market. In 1913 German organic chemist Friedrich Bergius (1884–1949) developed a high-pressure hydrogenation (hydrocracking) process that transformed heavy oil and oil residues into lighter oils, again boosting production for petrol. Later the German company IG Farben developed a similar high-pressure process, having acquired the patent for the Bergius process.

Bergius received the Nobel Prize in Chemistry for his work in 1931.

Two researchers in Germany, Franz Fischer and Hans Tropsch, created the *Fischer–Tropsch* process, to produce synthetic gasoline from coal in 1925. This involved the hydrogenation of carbon monoxide from combining either coke or crushed coal with heavy oil and steam, while exposing the mixture to a metal catalyst. The process played a critical role in meeting the increasing demand for gasoline and eased Germany's shortages during World War II. ICI built the first UK plant at Billingham to produce gasoline from coal in 1935. After the war, Germany was banned from hydrogenating coal; but the Soviet Union and South Africa long continued to use the Fischer–Tropsch process to make synthetic fuels from coal.

The next most important advance in hydrocarbon processing technology was the *catalytic cracking process* created by Frenchman Eugene Houdry in the 1930s.

Standard Oil Company's cracker Baton Rouge (1942)

Pioneer

Eugene Houdry (1892–1962) had graduated in mechanical engineering before joining the family metalworking business in 1911. He served in the French tank corps during World War I, and afterwards, while pursuing his interest in car racing during a trip to the US, he visited Ford's Motor

Company factory and attended the Indianapolis 500 race. This led him to take an interest in improving fuels for racing. As France had little petroleum itself, Houdry, like many other chemists, searched for a method to make gasoline from its plentiful stocks of lignite, a low-quality, brownish-black coal. The French government supported him to develop a successful demonstration plant, but it proved expensive and the government withdrew its support.

Undaunted, Houdry continued his research, discovering that a porous, silica-alumina material called fuller's earth (activated clay) worked well as a catalyst for converting the oil derived from lignite into a gasoline-like substance. Houdry then changed his feedstock to heavy liquid tars. By 1930 he had produced samples of gasoline that showed promise as a motor fuel. His process was the first to make catalytic cracking commercially viable. Unlike modern crackers, Houdry's unit employed a semi-batch process with a fixed bed of catalyst.

Houdry's advance interested many oil companies, including the Vacuum Oil Company (later the Socony-Vacuum Corporation) of Paulsboro in New Jersey, which invited him to the US to test his process on a larger scale. Houdry successfully scaled up the plant from 20 barrels a day to 200, but found the catalysts used to crack the hydrocarbon chains became quickly deactivated by the formation of a layer of coke on the catalyst surface. It was possible to remove the coke by heating and burning it off, but this took several minutes, so a batch plant would only be active for a very short fraction of its operating life and spend the rest of the time regenerating, making the cracking process uneconomical. His breakthrough was to devise the first practical way of regenerating the catalyst without shutting down production entirely. He employed multiple reactors, so at any given time some of the reactors were in operation while the others were regenerating.

While many oil companies were impressed with his method, it took a meeting with Sun Oil's president, J. Howard Pew, to obtain funding to further develop the process. Houdry later referred to this meeting in 1936 as *'the most beautiful day of my life'*.

Socony-Vacuum transformed a thermal-cracking unit at Paulsboro to use the *Houdry process*. In just under a year, it was processing 15,000 petroleum barrels a day. Houdry had developed catalytic cracking without the use of high pressure, producing more gasoline from oil that burned more efficiently in high-compression engines. The first commercial Houdry unit was built at Sun Oil's Marcus Hook oil refinery in Pennsylvania in 1937. The new units for gasoline and aviation fuel production proved vital for Allied success in World War II. During the first two years of the war, 90% of aviation gasoline came from 14 Houdry fixed-bed catalytic plants, which gave British planes a decisive edge over those of Germany during the Battle of Britain. The higher-octane rating provided up to 30% increase in power during take-off and climbing compared with other available fuels. Increasing demand for high-octane gasoline required for the ever more powerful aero-engines drove rapid deployment of the catalytic cracking process during the war.

Houdry's basic technology remains important in oil processing. Soon there were many further advances in catalytic cracking processes as chemical engineers made developments in reactor design driven by energy conservation and process kinetics.

In 1941 the introduction of *Thermafor* catalytic cracking (TCC) made the process semi-continuous by placing the catalyst in a bucket on a conveyor belt linking regeneration and cracker. This integrated endothermic cracking and exothermic catalyst regeneration operations. The catalysts used were synthetic alumina or silica beads that had more homogeneous and consistent activity than natural minerals. In this process, catalyst particles and the feedstock are introduced from the top of the reactor, the catalyst particles move downward with gravity and the cracking reactions take place on the catalyst surface. Steam is injected from the bottom of the reactor to carry the cracking products to the fractionator for recovery. As the particles move down the reactor, they are deactivated by coke build-up on the active sites. The deactivated catalyst removed from the bottom of the reactor is sent to a regenerator unit where the coke on the catalyst surface is burned off and the heated catalyst particles are recycled back to the top of the reactors by bucket elevators. The hot catalyst particles provide most of the heat necessary for the cracking reactions in the reactor. Although the thermal efficiency of TCC is higher than that of the Houdry process, there is still a significant amount of heat loss during the transport of heated catalyst particles by bucket elevators.

The successful Houdry process was breaking heavy long-chain hydrocarbons into shorter and more usable fractions. It attracted competitors to enter the market. Chemical engineers at Standard Oil attempted to develop a continuous process that would maximize the contact area between the catalyst and the crude oil before moving the deactivated catalyst from the cracker to the regeneration chamber without shutting down the cracker, nor allowing the air from the regeneration chamber to meet the hydrocarbons in the cracker. While screw conveyors could achieve this in small units, they were difficult to scale up because of their tendency to clog up and experience excessive wear. Warren Lewis and Edwin Gilliland of MIT suggested that a very fine powdered zeolite catalyst might behave like a liquid if a gas was blown through it at low velocity. The catalyst is mixed with vaporised oil flowing at low velocity. The vapour/solid mixture flows through the reactor like a liquid while the cracking reaction takes place. The cracked vapours are then separated from the spent catalyst with the aid of a steam stripper, and the catalyst transferred to the catalyst regenerator where the coke is burned off before the catalyst is recycled. The development of the first-ever FCC (fluidized catalytic cracking) unit required solving many engineering problems. They designed the moving-bed catalytic converter, in which the catalyst was itself circulated between the enormous reactor and regenerator vessels.

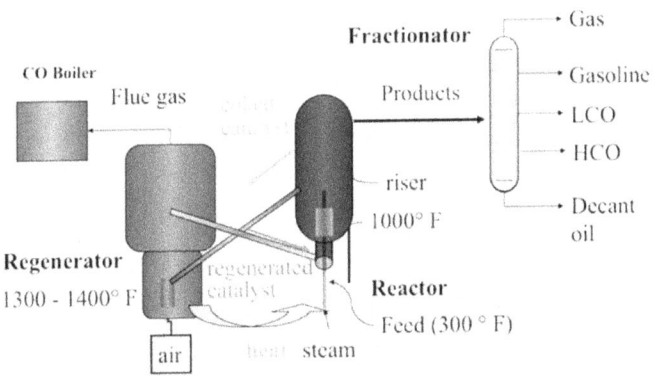

Process for Fluid Catalytic Cracking (FCC)

Gilliland wrote: *'There were many problems to be solved in going to the tremendous-sized reactors needed... These were chemical reactors much larger than had ever been built for any reaction and they used a catalyst that had to be regenerated every few minutes.'*

The engineering contractors MW Kellogg built a 100-barrel per day pilot plant, which started up at Standard Oil's Baton Rouge refinery in May 1940. The first full-scale plant with an initial capacity of 13,000 barrels of oil per day followed one year later, then more new plants of an improved design that needed less steel to build were started up in 1942 and 1943. It was evidence of the success of the FCC process that both original units continued in operation at Baton Rouge, Louisiana, for over 50 years.

The need to conserve materials such as steel during wartime favoured fluid-bed catalytic crackers over fixed-bed units, which were less economical. Soaring demand for high-octane aviation fuel and the butadiene needed to produce synthetic rubber to advance the war in Europe meant that by 1945 over 30 more FCC units had been built. There are some 400 FCC units in operation around the world today and all of them can trace their ancestry back to the original units in Baton Rouge. FCC became the most widely used process because of its thermal efficiency and the high product selectivity achieved, particularly after the introduction of crystalline zeolites catalysts in the 1960s.

These advances in hydrocarbon processing technology ushered in the new era of petrochemicals. The increased need for more light hydrocarbons resulted from rapid growth in private car ownership, commercial aviation and the invention of many new materials. Chemical engineers and chemists had found new uses for the new feedstocks in hundreds of industries and created many new and exciting products such as nylon, polyester fabrics, PVC and polyurethane plastics.

Future transition

Modern hydrocarbon processing advanced with catalytic processes and fluid catalytic cracking, which are now considered some of the most influential chemical engineering achievements of the 20th century. Many more innovations followed in hydrocarbon processing that made oil vital in every aspect of modern life. As the world today uses massive amounts of energy from fossil fuels transitioning to alternatives is not straightforward, as oil and natural gas have provided an optimal combination of energy density, safety, price, ease of storage, and transportability.

Hydrocarbon processing is now being transformed by actions to address global warming following a legally binding international treaty agreed in 2015. This aims to limit global warming to below 2 °C by reducing greenhouse gas emissions to achieve net zero while building up resilience to the impacts of climate change (see Z). As an example, the UK government has set an ambitious target for 2050 of reducing the emissions from air travel by using more sustainable aviation fuels (SAF). These jet fuels can be waste-derived biofuels, recycled carbon fuels (making use of unrecyclable plastic and waste industrial products) or innovative new processes. Power-to-Liquids (PtL) technology uses renewables-powered electrolysis to produce hydrogen with a reverse water gas shift coupled with Fischer–Tropsch synthesis fed with carbon dioxide captured from air. These can reduce the net greenhouse gas emissions of aircraft by up to 90%, although they are currently markedly more expensive than fossil jet fuel.

In the future, new chemical technologies such as these will be vital; although to maintain present standards of living during the 'energy transition' some fossil fuels are still required. Oil will remain an important raw material feedstock for producing vital materials such as plastics for the foreseeable future and gas for baseload power generation. Large energy businesses will increasingly focus on the production of low-carbon energy from renewables and on carbon dioxide capture, hydrogen gas production and storage, using geothermal sources and investing in natural carbon sinks.

I: Iron and steel

Steel bridge on Alaska Highway Canada

Backbone of the Industrial Revolution

Iron is one of the most important metals in history as workable iron ores have been used for more than 3,000 years in most parts of the world. The chemical technologies for iron making depend on removing unwanted elements from the ores and controlling those that are beneficial. In 1722 French physicist René de Réaumur first investigated how different types of iron such as wrought iron, cast iron, pig iron and steel were distinguished by the amount of carbon they contained.

Iron is the second most common metal in the earth's crust (about 5%) and much of the core is thought to be composed of iron. In its pure form it is a brittle, soft metal and rapidly corrodes in exposure to moist air and high temperatures. On heating, pure iron melts at a temperature of 1,538 °C. Iron forms an important role in living organisms – in animals it is a component of haemoglobin, the protein in blood that carries oxygen in the body, and in plants it is essential for photosynthesis.

The Industrial Revolution from the 18th century was to rely extensively on new iron-based technologies, ushering in an era of using iron and then steel, giving rise to a world driven by new machinery, pumps and steam engines, the appliance of new agricultural tools, transportation by sea, rail and road, and the construction of myriad buildings, civil structures and factories. Steel is an alloy of iron and carbon with improved strength and fracture resistance compared with other forms of iron, combining the toughness of wrought iron and the hardness of cast iron. The advent of modern steel followed from understanding the impact of varying amounts of carbon (often between 0.1% and 2%) and adding trace metals to produce steel alloys with other useful qualities. About 90% of all iron is now used to produce this extremely strong, machinable, corrosion-resistant and versatile material.

In the mid-19th century, fine craftsmanship of steel was further developed, but more important was the discovery of new processes and techniques to economically mass produce it. The chemical technology invented by British engineer Henry Bessemer made steel cheap and prevalent in a wide variety of uses.

History

The metallurgy of iron and its alloys is revealed in ancient history. Some of the earliest surviving iron artefacts dating from the 4th millennium BC in Egypt are made from rare iron-nickel that had fallen to earth in meteorites. The process of extracting iron from its naturally occurring ore by heating with a reducing agent such as carbon is known as *smelting*. It is not known where the smelting of iron from ores first began, but iron was being produced from ores in Mesopotamia in the Middle East, India and possibly in sub-Saharan Africa. By about 1500 BC, many non-meteoritic, smelted iron objects appeared in these regions. The iron was originally smelted in bloomeries, furnaces using bellows to force air through a pile of iron ore and burning charcoal. The carbon monoxide produced by the charcoal reduced the iron oxide in the ore to metallic iron. The bloomery furnace, however, was not hot enough to melt the iron, so the metal collected in the bottom of the furnace as a spongy mass – the 'bloom'. Workers had to repeatedly beat and fold it to force out the molten slag, a laborious, time-consuming process that produced wrought or worked iron, a malleable but soft iron alloy with a low carbon content (less than 0.10%). The spread of wrought iron defined the *Iron Age* as the period from about 1200 BC when bronze weapons and tools were replaced with those of iron and steel. That transition happened at different times in different places with the spread of iron technology.

Ironsmiths discovered that wrought iron could be turned into a much harder product by heating the finished piece in a bed of charcoal and then quenching it in water or oil. This was *carburization*, the process of adding carbon to wrought iron. While the iron bloom contained some carbon, the subsequent hot-working oxidized most of it. This process turned the outer layers of the piece into steel alloy and iron carbides, with an inner core of less brittle iron.

The earliest known production of steel occurred in Anatolia (modern Turkey). Around 500 BC, high-quality steel was produced in southern India by the 'crucible' technique. High-purity wrought iron, charcoal and glass were mixed in a crucible and heated until the iron melted and absorbed the carbon. The manufacture of *Wootz* steel, which has a unique structure and a carbon content between pig iron and wrought irons, was developed in India and Sri Lanka from around 300 BC, and it became known for its durability and ability to hold a sharp edge.

In China, iron was widely used for most tools and weapons by 300 BC. Chinese metallurgists in the Han dynasty (206 BC–220 AD) created steel by melting together wrought iron with cast iron to produce a carbon-intermediate steel alloy. Their technology spread to the Islamic world and then onto Northern Europe. By the 1st century BC, Nordic steels were highly sought after in the Roman empire, which controlled over 80,000 tonnes of iron production.

By 500–400 BC, use of iron artefacts had been adopted across the British Isles. The population of Britain grew substantially during the Iron Age, exceeding one million, as farming techniques improved and iron-tipped ploughshares made the cultivation of heavy clay soils possible. New methods of making pots, working wood and grinding grain were all facilitated by iron. The Anglo-Saxons refined the mastery of steelmaking to produce sharp-cutting swords. In the mid-medieval period, iron technology included the famous *Damascus* steel in the Middle East for sword making. This used the crucible steel method, based on the earlier Wootz steel. Modern analysis suggests that special nanostructures that give this innovative steel its distinctive properties are a result of the forging process.

In Europe for many centuries, iron was still being made by the working of iron blooms into wrought iron. A way was found of producing wrought iron from pig iron by decarbonization in a 'finery' forge. Cast iron is a brittle metal containing about 2% carbon. By the late 14th century, the market for cast iron began to grow from demand for cannonballs and cooking wares. A new type of furnace called the 'blast furnace' was devised in Liege, Belgium, to produce a form of cast iron by smelting the iron ore with charcoal. In the UK the short availability of charcoal sourced from tree wood was limiting the expansion of iron production, so that Britain became increasingly dependent on iron imports from Sweden and Russia. This led to the development of smelting with coal (or its derivative coke).

In the early 17th century, ironworkers in Bavaria developed the 'cementation' process for carburizing wrought iron bars. This process was introduced to England in about 1614 and used to produce steel by Sir Basil Brooke at Coalbrookdale in Shropshire, England. It was thought the best steel was produced when using 'oregrounds', iron bars shipped from Sweden. In 1740 Benjamin Huntsman, an English inventor, developed the crucible process for casting steel. A crucible was used to melt the blister steel, made by the cementation process, rather than having been forged. The resulting crucible steel, made into ingots, was more homogeneous than blister steel. Most previous furnaces could not reach high enough temperatures to melt the steel.

In 1707 Abraham Darby ('The Elder') had patented a method of making cast-iron pots. His pots were thinner and hence cheaper than those of his rivals. Needing a larger supply of pig iron, he leased the blast furnace at

Coalbrookdale in 1709. There, he established the first business making cast iron using coke. This reduced the price of iron, although the coke-fuelled pig iron could not initially be converted to bar iron by the existing finery forge methods due to sulfur impurities from the coke until increased furnace temperatures allowed the sulfur to be removed. In 1755 his son, also Abraham Darby ('The Younger') opened a new coke furnace at Horsehay in Shropshire. An economically viable means of converting pig iron to bar iron began to be devised as the finery forge process was replaced by the 'puddling' process developed by Henry Cort in Hampshire in 1784. Cort's process consisted of stirring molten pig iron in a reverberatory furnace in an oxidizing atmosphere, thus decarburizing it, to produce wrought iron. Although it was not the first process to produce bar iron without charcoal, puddling was by far the most successful, and it replaced the earlier potting and stamping processes, as well as the much older charcoal finery and bloomery processes.

Many more process innovations followed, enabling a great expansion of iron production to take place in Britain, which was essential to the unfolding Industrial Revolution. It was also found possible to produce steel by stopping the puddling process before decarburization was complete. By the 1830s the Dowlais Ironworks at Merthyr Tydfil in South Wales was the largest steel plant in the world. As the puddling process could not be scaled up, engineers were using cheaper cast iron for making bridges, ships and railway rails. The introduction of a new method of mass-producing cheap steel from Henry Bessemer at his steelworks in Sheffield, England, in 1855 would revolutionize the use, scale and economics of steel production.

Pioneer

Henry Bessemer (1813–1898) was born in Hertfordshire. His father had developed a machine to produce steel dies at the Paris mint. Henry became interested in metallurgy at his father's foundry. He made a machine producing very fine brass powder for making a cost effective 'gold' paint; a typesetting machine; a way of compressing graphite powder for pencils; a screw extruder for extracting sugar from sugar cane; and a process for producing a continuous strip of plate glass. It was his experience using furnaces that proved to be invaluable later.

Bessemer's interest in steelmaking began at the outbreak of the Crimean War in 1853. He later said his work was inspired by a conversation with Napoleon III about the steel required for better artillery. He produced a grooved artillery projectile for the army that could be fired from smooth-bore guns. However, the French army felt them too heavy to be fired safely from their brittle cast-iron guns. Bessemer had realized that it would be better to use steel but it would have been too costly. He decided to find a way of making a cheaper steel from brittle pig iron, which contains significant (3–4%) amounts of carbon. Initially he used a reverberatory furnace which separates the metal from the combustion fuel. In one experiment, some pieces of pig iron lying at one side in the stream of hot air refused to melt, even when he raised the temperature. Bessemer had discovered that very hot air alone had turned the outside of the ingots into steel shells. He later wrote: *'I became convinced that if air could be brought into contact with a sufficiently extensive surface of molten crude iron, it would rapidly convert it into malleable iron.'* Bessemer adapted the furnace to force cold air straight through the molten iron, even though his workers told him it was a foolish plan as the air would cool the iron. He was proved right as the airflow caused silicon and carbon to oxidize in an exothermic reaction, which greatly raised the temperature of the process rather than cooling it.

The first time he ran the new process, Bessemer observed how violent it was: *'All went on quietly for about ten minutes ... sparks, accompanied by hot gases, ascended through the opening on the top of the converter.... But soon after, a rapid change took place; in fact, the silicon had been quietly consumed, and the oxygen, next uniting with the carbon, sent up an ever-increasing stream of sparks and a voluminous white flame followed by a succession of mild explosions. However, after ten minutes the eruption ceased, the flame died down, and the process was complete. On tapping the converter into a shallow pan to form the metal into an ingot, it was found to be wholly decarburized malleable iron.'*

All attempts to make the reaction less violent failed but Bessemer concluded that speed, extreme heat and the eruptions were all the necessary for successful steel production. So, he focused on designing a reaction vessel that would safely contain the process. He created an egg-shaped reaction vessel with plenty of headroom for molten metal to spark and erupt without leaving the converter and gave it a special outlet or 'mouth'. The converter was pivoted on trunnions so that it could be tilted to receive its 15-tonne charge of molten pig iron from a blast furnace and pour the molten steel product into moulds.

He patented the process in 1855, although initially it failed to produce the high-quality steel expected as excess oxygen left the steel too brittle and it was difficult to retain the right quantity of carbon in the steel (0.2–2.1%). After the carbon content in the melt had dropped to the desired level, the air was cut off. A better solution was proposed by the metallurgist Robert Mushet, who suggested burning off all the carbon and then adding a precise quantity of Spiegeleisen, an alloy of iron, carbon and manganese. This made it possible to ensure the steel contained the right quantity of carbon, while the manganese removed excess oxygen from the steel. The improved Bessemer process was first licensed in 1865 to the Dowlais Iron Company.

While the typical charge for a puddling furnace was 400 kg, a Bessemer converter could convert a 25-tonne batch of pig iron to steel in half an hour. Bessemer made 'mild steel' which had many advantages over wrought iron and it allowed much larger rolled steel items to be produced. As a result of the vast increase in scale and speed of production the new Bessemer process decreased the cost of steel by a factor of 10 to typically less than £6 per tonne. This suddenly made possible the mass production of strong steel girders for bridges, buildings, railways, large ships and skyscrapers.

Bessemer Converter at Kelham Island Museum, Sheffield

Bessemer founded the Henry Bessemer and Co. in Sheffield, a town that became known worldwide for its steelmaking expertise, to commercialize the process. In the US, the earliest Bessemer converters were built near Detroit, providing the starting point for the city's later rise as the centre of the American car industry.

Bessemer's original process was unable to remove excess phosphorus, which also caused the steel to be brittle. This problem was eventually solved by Sidney Thomas in 1879, who introduced a refractory lining of clay or limestone to the converter. By the time he died in 1898, the global annual production of Bessemer steel ran to £80 million (over £1 billion today) – an astronomical sum for the time. Henry Bessemer had become very wealthy and saw the huge impact of his innovation in chemical technology. The Bessemer process remained in use for over 100 years with the final converter only ceasing production in 1968.

Steel for the future

Bessemer's chemical technology led to the widespread availability of inexpensive steel for building the modern world. Steel became the literal backbone of the global economy and its infrastructure. The Bessemer process was eventually made obsolete by the process of basic oxygen steelmaking (BOS), invented in 1948 by Swiss engineer Robert Durrer (1890–1978), in which carbon-rich molten pig iron is made into steel. Blowing pure oxygen through molten pig iron lowers the carbon content of the alloy and changes it into low-carbon steel. The process is 'basic' because alkali fluxes of burnt lime or dolomite are added to promote the removal of impurities and protect the lining of the converter.

Global steel production grew from 190 million tonnes in 1950 to over 1,900 million tonnes in 2020. In this period the energy cost of steel has continued to fall at an annual rate of 1.7%. Over half of all steel is made in China. Significantly, the steelmaking industry is still one of the most energy and carbon emission intensive. On average, every tonne of steel produced results in the emission of 1.85 tonnes of carbon dioxide, which amounts to about 8% of all global greenhouse gas emissions.

Many modern commercial plants use the electric arc furnace (EAF) for making steel. The basic oxygen process using coal emits four times more greenhouse gases than the electric furnaces as these are fed by recycled steel. The EAF uses scrap steel or direct reduced iron (DRI), which is produced from the reduction of iron ore by a reducing gas mixture of hydrogen and carbon monoxide. Steel made using DRI requires significantly less fuel as a traditional blast furnace is not needed. The UK has closed its ageing blast furnaces and now imports half of its steel from Asia, while the remainder is steel produced in EAFs fed from scrap metal as part of the move to a net zero supply chain. Apart from maximizing use of EAFs powered by renewable energy and using more recycling to mitigate increased use of new steel, the steel industry is currently developing several new chemical technologies for decarbonization including using 'green' hydrogen as a reducing agent (generated from renewable energy) or future employing of carbon capture and storage. Modern steels will be essential to build the new homes, cars, trains, solar arrays and windfarms in a future decarbonized world. Even a single large renewable energy wind turbine requires up to 1,000 tonnes of steel to be used in its construction. It likely that 'greener' steel production is going to be more expensive than the steel production today.

J: Joliot-Curie artificial radioisotopes

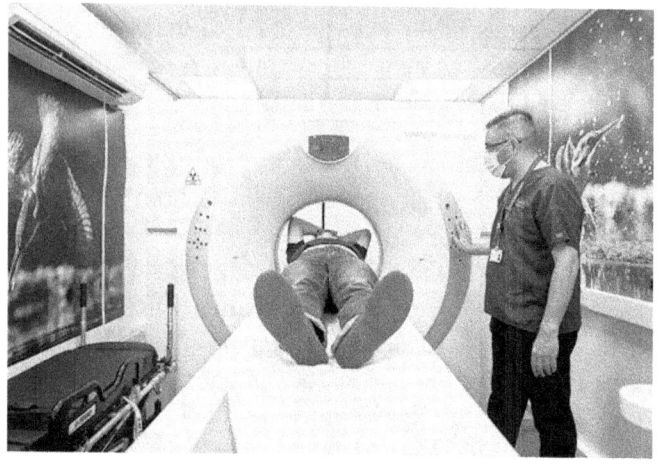

Scanner screening for lung cancer

Life-saving radiation

Artificial radioisotopes were discovered by Irène Joliot-Curie with her husband Frédéric Joliot. These isotope sources of artificial radioactivity have found many essential applications in medicine, scientific research, engineering diagnostics and measurement.

Radioactivity (or radioactive decay) is the process by which the nucleus of an unstable atom loses energy by emitting radiation, in the form of sub-atomic matter such as alpha particles (α-decay), beta particles (β-decay) or gamma rays (γ-decay). Radioactive decay is a random process at the level of single atoms, but the overall rate is expressed as a 'half-life' ranging from nearly instantaneous to millions of years. Although radioactivity is observed as a natural occurring process, the Joliot-Curies found that it can also be artificially induced by bombarding atoms of a specific element with radiating particles, thus creating new atoms. Previously the only way to obtain radioactive elements was to painstakingly extract them (as Irène's parents had), from their natural ores at considerable cost.

Ernest Rutherford (1871–1937), a New Zealand-born professor of physics at Cambridge, was an early pioneer of nuclear physics. He reported the existence of alpha and beta rays in uranium radiation, then in 1902 he proposed that radioactivity results from the disintegration of atoms, finding that the nuclei of certain light elements, such as nitrogen, could be 'disintegrated' by the impact of energetic alpha particles coming from a radioactive source. In 1910 his investigations into the scattering of alpha rays and the nature of the inner structure of the atom led to his concept of the nucleus. He showed that practically the whole mass of the atom and all its positive charge is concentrated in a minute space at its centre.

In the 1930s the Joliot-Curies discovered that although atoms appear to be stable, they can be transformed into new atoms with different chemical properties. Radioactive isotopes can be made by bombarding elements with α-particles, deuterons, protons, electrons, neutrons and even high-energy X-rays.

Pioneers

Irène Joliot-Curie (1897–1956) had very famous parents, who had themselves pioneered research into radioactivity and chemistry. Her Polish-French mother was Marie Skłodowska (1867- 1934), a double Nobel laureate, known for her pioneering studies of radioactivity.

Marie had studied chemistry at the Sorbonne in Paris, where she met and married Pierre Curie, a physicist who had achieved recognition for his work on the piezoelectric effect. For her thesis she chose to work in a field recently initiated by Wilhelm Roentgen's discovery of X-rays and Becquerel's observation of the mysterious power of uranium salts to expose completely covered photographic film. Marie Curie convinced her husband to join her in isolating a 'substance' that she was the first to refer to as 'radioactive'. Radioactive isotopes have the same chemical properties as stable isotopes of the same element, but they emit radiation, which can be detected. In 1898, after laboriously isolating various substances by successive chemical reactions and crystallizations, they then tested for their ability to ionize air. The Curies had discovered polonium, and then radium salts weighing about 0.1 g that had been derived from a massive quantity of crude uranium ore (pitch blend).

Marie received her first Nobel Prize in 1903 for physics, shared with her husband Pierre Curie and Henri Becquerel, for the discovery of the phenomenon of radioactivity, and her second in 1911 in chemistry, for the discovery of the radioactive elements polonium and radium. Radium institutes were established for her in France and in Poland to pursue the scientific and medical uses of radioactivity. During World War I Curie organized a field system of portable X-ray machines to help in treating wounded French soldiers.

Daughter, Irène, worked with her mother in the radiography service and then assisted her at the Radium Institute in Paris, while completing her doctorate. Irène married Frédéric Joliot (1900–1958), a young physicist who worked for her mother.

In their first experiments Irène and Frédéric Joliot-Curie made use of the large supply of polonium generated by her parents' work. The polonium emitted alpha particles, which they used to bombard different elements. In 1933 they bombarded an aluminium plate. When they removed the alpha particles, it appeared that the aluminium plate emitted radiation with a half-life of 3 minutes. Their explanation was that the bombardment had resulted in a nuclear reaction as the alpha particle had penetrated the aluminium nucleus and changed it into one of phosphorus by emitting a neutron. The new phosphorus isotope (P-30) resulting was responsible for the radiation. This nuclear reaction may be represented as 27Al + 4He (alpha particle) $\rightarrow$ 30P + 1n(neutron). The Joliot-Curies were awarded the Nobel Prize in Chemistry in 1935 for having created new artificial radioactive elements by the bombardment of alpha particles on various light elements.

Unfortunately, like her mother, Irène was to die of leukaemia induced by long exposures to radioactive materials in her work without the benefits of modern health and safety precautions.

Irène and Frédéric Joliot-Curie in their laboratory (1935)

Benefiting from radiation

With their discovery of artificial radioactivity, radioactive atoms could be prepared relatively inexpensively, leading to advances in nuclear physics, medicine and many other applications. They are typically used to follow or monitor the paths of chemical reactions or to determine how a substance is distributed within an organism, to measure machine wear and metal thicknesses, to analyse geological formations and most importantly in vital life-saving medical treatments.

Radioactive isotopes have the same chemical properties as stable isotopes of the same element but emit radiation, which can be detected when replacing it with a radioisotope in a compound, so it is possible to trace them by monitoring their radioactive emissions. In this way they are used to study the mechanisms of chemical reactions taking place in plants and animals. Fertilizers can be 'labelled' to show nutrient uptake by plants. The uptake of phosphorus from a fertilizer can be traced in the plant by using a fertilizer containing radioisotope P-32. A radioactive carbon (C-14) was used to determine the working of the complex photosynthesis process.

It is also possible for a very small amount of a radioactive isotope to be attached to an object so it can readily be detected by picking up the radiation emitted. A joint in a buried pipe or cable can be marked, or the efficiency of a mixing process can be followed by adding a radioactive isotope to one of the ingredients before mixing and then observing the level of radiation throughout the final mixture. The flow of a material, such as molten glass through a furnace, powder through a drier, or gases in a ventilation system, can also be measured by this technique.

Radioactive tracers are used in many medical applications, including both diagnosis and treatment. Radioisotopes have revolutionized medical practice, with millions of nuclear medicine procedures and tests being performed annually. An example of a radioactive tracer used in medicine is the metal isotope technetium-99 (Tc-99) for locating damaged tissues in the body. The isotope thallium-201 (Tl-201) becomes concentrated in healthy heart tissue so is used to investigate heart function and blood flow, while iodine-131 (I-131) has been used in the treatment of thyroid disease. When used for the treatment of cancer in radiation therapy, high doses of radiation kill cancer cells and shrink tumours. Carefully directed high-energy radiation is used to damage the deoxyribonucleic acid (DNA) of cancer cells to destroy them or prevent them from dividing. A cancer patient may receive an external beam of radiation therapy delivered from outside the body, or internal radiation therapy (brachytherapy) from a radioactive substance that has been introduced into the body.

In engineering, synthetic isotopes are used to assess the wear inside an engine or machine by making various moving parts radioactive and measuring the resulting radioactivity in metal particles within the circulating oil. Radioactive isotopes can also be very useful when measuring or controlling thicknesses of materials. The amount of radiation passing through a material will decrease as the material gets thicker. In manufacturing processes such measurements can be made without touching the material concerned, so that they can be applied, for example, to hot sheet steel coming through a high-speed rolling mill, or controlling the thickness of plastic sheeting. It is possible to monitor the thickness of a coating of one metal on another or the level to which a container or tank is filled with a liquid or solid. Structural flaws in metals can be located using high-energy gamma rays from a cobalt-60 isotope in high-pressure pipes and vessels.

Minute quantities of radioactive isotopes are used in ionization-type smoke detectors. The alpha emissions from the americium isotope (Am-241) ionize the air between two electrode plates in the detector's chamber creating a small electric current. When smoke particles from a fire enter the chamber, the movement of the ions is impeded, reducing the conductivity of the air. This causes a drop in the current, which activates a loud alarm. In food preservation, since ionizing radiation can kill bacteria, moulds and insect pests and reduce the ripening and spoiling of fruits, its use may be compared to pasteurization. It has sometimes been referred to as 'cold pasteurization' as no heating is required.

As a result of the pioneering work of the Joliot-Curies on artificial radioactivity, a whole new chemical technology was developed to manufacture artificial radioisotopes that are exploited in many useful and life-saving applications. Many millions of people every year depend upon medical imaging to diagnose disease and detect injury, and thousands more rely on radiation therapy to treat and cure their cancers. Over a thousand different artificially created radioactive nuclides have been produced.

K: *Kryptonite* stands aside for graphene

A real 'super material'

Kryptonite is the fictious green crystalline 'super material' that appears in the US *Superman* comic stories, originating from Superman's home planet of Krypton. It emitted a peculiar radiation that was harmless to humans but very dangerous to the super-hero himself. Conveniently avoiding the need to find a technology starting with the letter K, here is not a historical chemical technology but one at the beginning of its existence. *Graphene*, in the real world, is an actual 'super material', although graphene's long list of miraculous properties makes it seem as though it too is another fiction; it could have very real and wide-ranging implications in the future. Graphene was first created by Andre Geim working with Konstantin Novoselov in 2004. The new material is interesting to quantum physicists but could lead to the creation of a range of new materials and revolutionary applications.

Graphene is a special form of elemental carbon. A material that is stronger than steel, thinner than paper and a good conductor. Materials have shaped the growth and progress of the modern technological world. Graphene could be part of many new future technologies, in much the same way as bronze and iron were so crucial to the spread of ancient civilization, or the rise of steel in the Industrial Revolution helped create modern engineering, then silicon semiconductors enabled the computer era and the spread of information technology.

Carbon is a fascinating element and the basis of all life on earth. It can exist in several different forms. The most common form is graphite, which consists of stacked sheets of carbon with a hexagonal structure – the soft, flaky material used in pencils. Graphite is an *allotrope* of carbon, meaning it possesses the same atoms but they are bonded in a different way, giving the material different properties. Another allotrope is diamond, formed under very high temperatures and pressures, which is a giant network form of carbon. Diamonds are incredibly strong, but graphite is very brittle. Other forms of molecular carbon include the *fullerenes* and carbon *nanotubes*.

The simplest way to describe graphene is that it is a single thin layer of graphite with its atoms in a hexagonal arrangement. Graphene, being only one atom thick in a flat atomic lattice, confers exceptional properties. Graphene was the first two-dimensional (2D) material ever discovered; the thinnest and the strongest, with properties that may give it many potential future applications.

Pioneers

This unique material was first created by Andre Geim working with Konstantin Novoselov. Geim was born in Russian in 1958 and later became a Dutch-British physicist working in the School of Physics and Astronomy at the University of Manchester. At university he studied solid-state physics, obtaining a master's degree from Moscow Institute of Physics and Technology in 1982 and then a postgraduate degree in metal physics from the Institute of Solid State Physics (ISSP) at the Russian Academy of Sciences.

Geim's co-discoverer, Konstantin Novoselov, was also born in Russia (in 1974) but is now a Russian-British physicist. He also graduated from the Moscow Institute in 1997, and he was awarded a doctorate in 2004 from the Radboud University of Nijmegen for work that was supervised by Andre Geim. In 2001 Geim became a professor of physics at the University of Manchester. It was while working at the University of Manchester that they made their important contribution to chemical technology.

Graphite had been used since prehistoric times in art and for decorating pottery, then as a refractory material to line moulds to make smoother cannonballs that could be fired farther, improving the effectiveness of the British Navy. During the 19th century, graphite was used to make stove polish, lubricants, paints, crucibles and foundry facings and for producing cheap pencils, which assisted the expansion of education. Its atomic structure is well understood, and for a long time scientists had postulated whether single layers of graphite could be isolated, slicing graphite down to a single, atom-thin sheet. Most believed it was impossible for such thin crystalline materials to be stable. It was well known that graphite consists of hexagonal carbon sheets that are stacked on top of each other, but it was believed that a single sheet could not be produced in an isolated form. It came as a surprise then, when Geim, Novoselov and some collaborators from the University of Manchester and the Institute for Microelectronics Technology in Russia presented their results on graphene.

When they isolated the first sample of graphene it might have been expected that they were using complex and expensive laboratory equipment, but the tools they used were very simple. The two professors frequently held informal 'Friday-night experiment' sessions where they would try out experiments that were not linked to their main research. In a very simple experiment, they removed some flakes of graphite from a lump of bulk graphite with adhesive tape. They noticed some flakes were thinner than others as they 'polished' a large block of graphite with the tape, and some flakes on the tape were exceptionally thin. Continuing to peel layer after layer from the flakes of graphite, they eventually produced a sample as thin as is possible. They had used a simple mechanical exfoliation method for extracting thin layers of graphite from a graphite crystal and then transferred these layers to a silicon substrate. By separating the graphite fragments repeatedly, they managed to create flakes that were just one atom thick. Their experiment had led to 'graphene' being isolated for the very first time. They were able to not only isolate but identify this new material. Although the scientific community was initially sceptical. They published their results in October 2004, fully describing their work including the microscopic characterization of graphene. It is a crystalline two-dimensional material, one million times thinner than a single human hair, and it was difficult to isolate sufficiently large individual sheets of graphene to be able to identify and characterize the material and access its unique two-dimensional properties. For this discovery they were jointly awarded the Nobel Prize for Physics in 2010.

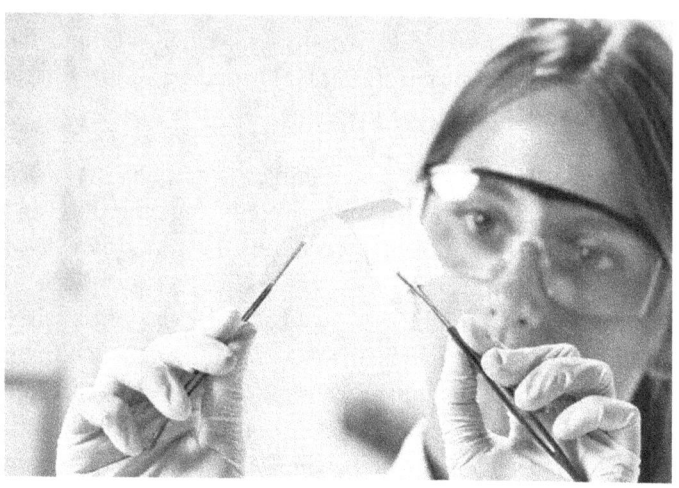

Graphene advanced composite material

The potential of graphene

After the discovery of graphene, Geim also developed a biomimetic adhesive, which is very good at sticking to surfaces and was nicknamed 'gecko tape' after the lizard that climbs walls and ceilings. They hoped that graphene and other two-dimensional crystals would change everyday life as much as plastics have done. Novoselov became a director of the National Graphene Institute in Manchester to research and exploit its potential.

Having unique properties makes it potentially useful in a vast number of products, processes and industries. Most importantly, graphene is many times stronger than steel, yet incredibly lightweight and flexible, electrically and thermally conductive but also transparent and seemingly impermeable to most gases and liquids. It has a tensile strength – the maximum stress that a material can withstand while being stretched before breaking – that is more than 100 times stronger than quality steel. As a conductor of electricity, it performs as well as copper. As a conductor of heat, it outperforms other materials, its thermal conductivity being 10 times greater than copper. It is almost completely transparent and colourless, absorbing only 2% of light yet so dense that not even helium can pass through it. Several other two-dimensional crystalline materials with unique properties have been made and analysed. After graphene, consisting of a single atomic layer of carbon atoms packed in a hexagonal (honeycomb) lattice, single layers of boron-nitride and molybdenum-disulfide were produced.

Graphene has the potential to create the next generation of electronics using faster semiconductors that exploit its properties of thinness and conductivity. Being just one atom thick and able to conduct electricity at room temperature, graphene semiconductors could be employed in computer chips. Researchers at the University of Manchester have already created the world's smallest transistor using graphene. Graphene-based chips would be substantially faster than those made of silicon transistors, resulting in more efficient computers. A graphene supercapacitor could provide massive amounts of power while using much less energy than conventional devices. The UK based company Paragraf is the first in the world to mass produce graphene-based electronic devices using standard semiconductor processes.

In addition, graphene has several remarkable mechanical properties. Since it is practically transparent and a good conductor, graphene is also suitable for producing transparent flexible touch screens for smartphones, light panels and solar cells. It is substantially stronger than steel but very stretchable. When mixed into plastics, graphene can turn them into electrical conductors while making them more heat-resistant and mechanically robust. This resilience can be utilized in new super-strong materials that are thin, elastic and lightweight.

In the manufacture of solar cells – compared with conventional cells – graphene can release multiple electrons for each photon that strikes it and so could be more efficient at converting solar energy for producing renewable energy. Graphene's photovoltaic properties also mean that it could be used to develop better image sensors for devices such as cameras. It was found that graphene demonstrates behaviour as a superconductor (a material with no electrical resistance) when paired with praseodymium cerium copper oxide (PCCO), but also acts as a superconductor alone, in the right configuration. When researchers stacked two slices of graphene, offset them by an angle of 1.1 degrees, a superconductor resulted. Most materials that display superconductivity only do so near the extremely low temperature of absolute zero (0 K or −273.15 °C). The graphene arrangement operated at some 1.7 degrees above absolute zero, so it will be a much easier material for studying unconventional superconductivity, currently an area of great uncertainty in physics. It may be possible to find a superconductor that operates at room temperature, which would bring down the cost of manufacturing by not requiring extreme cooling units.

More recently, using a near-industrial method, it has become possible to fabricate larger sheets (700 mm width) of graphene. The simplest and most effective way of harnessing the potential of graphene is to combine it with existing materials – so-called composite materials – for example paint. A graphene coating could signal the end of the deterioration of ships and cars through corrosion.

Sporting goods such as a graphene-enhanced tennis racket have already been produced while there is research into ways of enhancing sports equipment for skiing, cycling and even motor racing. A graphene-based composite aircraft wing could drastically decrease weight, reduce the detrimental effects of lightning strike damage and increase fuel efficiency and flight range. In the future, satellites, boats and cars could be manufactured of the composite materials containing graphene. Being very light would also lead to reduced fuel use.

As wearable technology has been growing in use, devices could be designed with integral graphene to fit to the body and accommodate various forms of exercise. Graphene's flexibility and microscopic width provide opportunities in biomedical research. Small machines and sensors could be made with graphene, capable of moving easily and harmlessly through the human body, analysing tissues. The body is based on carbon chemistry and so graphene can potentially be employed safely. Graphene's properties make it useful in ground-breaking applications such as targeted drug delivery, health-testing and 'smart' implants. To control spread of serious diseases like malaria, research has demonstrated that a graphene film on the skin not only blocked mosquitoes from biting but even deterred them from landing on skin in the first place.

Graphene could be utilized in new battery technologies to fully charge a smartphone in seconds or an electric car in minutes. It could dramatically increase the lifespan of a traditional lithium-ion battery, allowing more rapid charging and holding power for longer. A chemical sensor made using graphene in which every atom is exposed to the environment would allow it to sense very small changes in its surroundings.

Among future applications, graphene oxide could be used to create 'smart' food packaging products. This could dramatically cut down on unnecessary food wastage and simultaneously help prevent illnesses. Packaging that has been coated with graphene can detect atmospheric changes caused by decaying food. Graphene sensors could boost the effectiveness of monitoring vital crops in the agriculture industry. Farmers would be able to monitor the existence of any harmful gases that could impact upon crop fields. As graphene sensors are so sensitive it is feasible that they could determine the optimum areas for growing certain crops depending on atmospheric conditions.

Graphene's bonds are impermeable for nearly all gases and liquids, although water molecules are an exception, making it possible for the material to be employed in purifying water containing toxic contaminants (see W). This would have great benefits in the treatment of environmental waste, including nuclear waste and chemical pollution, which could be cleansed from water sources using graphene-based filters. Lockheed Martin developed a graphene composite filter called *Perforene* in 2013, which the company suggested could revolutionize the desalination process to produce fresh water. Most current desalination plants use reverse osmosis to filter salt out of seawater but require enormous amounts of energy to operate. Using graphene coatings on food and pharmaceutical packaging can stop the transfer of water and oxygen, keeping food and perishable goods fresher for longer.

The chemical technology of graphene is a completely new but revolutionary one. However, it is a long way from widespread commercial use, as it is still extremely expensive to produce in large quantities, which limits its use in any product that would demand mass production. In the future, graphene could be a disruptive chemical technology;

unlike its mythical predecessor *kryptonite*, it is a real super material that could create new markets and even replace existing technologies or materials.

L: Lithium-ion batteries

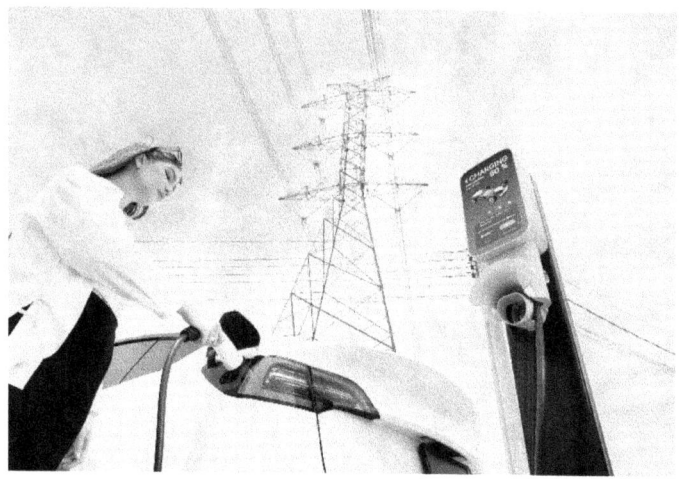

Recharging electric car (EV) from electrical power grid

Lifeblood of the modern world

Lithium-ion batteries are vital now in many ways, such as providing power to drive cars, for portable tools and mobile devices or as a back-up to mains electricity. They have become a truly ubiquitous chemical technology, often taken for granted, given their many benefits. Mobility is one of the most important attributes of modern batteries. Mobile technology would not have been possible without the crucial development of the lithium-ion battery (LIB) by the Oxford University chemist John Goodenough (1922–2023) with several other collaborators.

In 1980, while in the Inorganic Chemistry Department at Oxford, Professor Goodenough identified the cathode material that enabled development of the rechargeable LIB. Continuous development has ensured greater battery life, efficiency and versatility so that modern LIBs have been designed to fit into all manner of electrical equipment, from the tiny batteries used in watches, hearing aids, laptops and remote controls to slimline mobile phone batteries or larger batteries for powering cars. They are essential in transport, electronics, communication, computing and entertainment. Increasingly, batteries also have the potential to help reduce greenhouse gas emissions by efficiently storing electricity generated from renewable energy sources or providing a back-up power supply for critical infrastructure.

History

The term *battery* was first used for a group of connected electrical devices by Benjamin Franklin in 1748 when he described multiple connecting Leyden jars (devices for storing static electricity) by analogy to an artillery battery of military cannons. The first electrical battery able to generate an electrical current was invented in 1800 by Italian physicist Alessandro Volta (1745–1827). It was known as a *voltaic pile*: a stack of alternating copper and zinc plates, separated by brine-soaked paper disks, that could produce a steady current for a considerable length of time. Volta did not understand that the voltage was a result of electro-chemical reactions, but in producing a steady current his invention allowed for the more rigorous study of electricity, eventually laying the foundation for the revolutionary experiments of Michael Faraday (1791-1867) in electromagnetism. Over the next century various scientists worked to improve the effectiveness and life of batteries, so they would produce more electrical current and last longer.

The Daniell cell, invented in 1836 by British chemist John Daniell, consisted of a copper pot filled with a copper sulfate solution, in which was immersed an unglazed earthenware container filled with sulfuric acid and a zinc electrode. It was a reliable source of electrical power, becoming widespread as a power source for electrical telegraph networks. But 'wet cells' using acidic liquid electrolytes were prone to leakage and spillage if not handled correctly. Many used glass jars to hold the electrolytes, which made them fragile, potentially hazardous and unsuitable for portable appliances. The invention in 1859 of the lead-acid battery by Gaston Planté became the first rechargeable battery. It was made from lead plates suspended in a sulfuric acid electrolyte. Until then, battery life was limited, as once all the chemical reactants within the cells had been spent it would remain permanently discharged. Lead-acid batteries were first used to power the lights of train carriages and then in early cars in the late 19th century.

The first 'dry cell' used zinc and carbon to generate electricity. It was to transform the utility of batteries and drove the development of more portable devices. Alkaline batteries – employing an alkaline rather than an acidic electrolyte – were created by both Waldemar Jungner in 1899 and, working independently, Thomas Edison in 1901. The modern alkaline dry battery using zinc-manganese dioxide chemistry was invented by the Canadian engineer Lewis Urry in the 1950s.

In these alkaline batteries, the negative electrode is zinc and the positive electrode is manganese dioxide [MnO_2]. The alkaline electrolyte of potassium hydroxide is not consumed during discharge. The chemical energy is stored mostly in the zinc metal, whose free energy is at least 225 kJ/mol higher (so less stable) than that of the three oxides in the reaction. When first introduced the zinc electrode of alkaline batteries had a surface film of harmful mercury amalgam to control electrolytic action at impurity sites, which would reduce shelf life and promote leakage, but for environmental reasons battery makers were later forced to reduce their toxic mercury content.

The next big innovation in the chemical technology of batteries was creation of the *lithium-ion cell*. It is LIBs that have allowed a massive growth in portable and mobile electrical technology.

Pioneers

British chemist Stanley Whittingham (born 1941) became Professor of Chemistry and Materials Science at Binghamton University, New York. He laid the foundations for the battery during the 1970s as he worked to develop methods that could lead to fossil fuel-free energy technologies. His research into superconductors led to the discovery of an extremely energy-rich material, which he used to create an innovative cathode in a lithium battery. It was made from titanium disulfide as it could intercalate (insert) lithium ions. The anode was partially made from metallic lithium, which has a strong drive to release electrons. The resulting battery produced just over 2 volts, but as metallic lithium is very reactive, the cell was too explosive to be viable.

The American physicist John Goodenough was born in 1922, had studied mathematics at Yale University and served during World War II as a meteorologist in the US Army. He received his doctorate in physics from the University of Chicago in 1952. He worked at the MIT and then in the Inorganic Chemistry Laboratory at Oxford University. At Oxford, Goodenough predicted that the cathode would have even greater potential if it was made using a metal oxide instead of a metal sulfide. After a systematic search, Goodenough and research fellow Koichi Mizushima began researching lithium batteries. By 1980 they had improved on Whittingham's design, replacing titanium disulfide with lithium cobalt oxide, and demonstrated that a lithium-based battery can produce as much as 4 volts, a much higher voltage than earlier lithium batteries. This was an important breakthrough and would lead to much more powerful batteries.

In 1986 Goodenough became a professor at the University of Texas in Austin and, while credited with inventing the lithium-ion rechargeable battery, it was other engineers who turned his work into a practical reality. Akira Yoshino (born 1948), who worked at the Japanese chemical company Asahi Kasei, started to investigate the lithium battery, culminating in the creation of the first workable version. Instead of using reactive lithium in the anode, he used petroleum coke, a carbon-based material that, like cobalt oxide, can trap or intercalate lithium ions. The resulting battery was lightweight, hardwearing and could be charged hundreds of times before performance deteriorated. A fundamental advantage of LIBs is that they are not based on chemical reactions breaking down the electrodes. Rather, they are based on the flow of lithium ions between the anode and cathode.

The company sought expertise from engineers at Sony led by battery engineer Yoshio Nishi, who by 1989 had transformed a crude prototype into a commercial product.

Yoshio Nishi, born in Nagoya, Japan, had studied solid physical chemistry in the engineering department at Keio University in Tokyo and graduated in 1966. He would eventually spend 20 years developing and refining LIBs, starting as general manager of the LIB development team during the crucial early development phase. Sony was keen to develop lithium-based rechargeable batteries as an alternative to nickel-cadmium batteries, because they promised a much greater energy density and would attract fewer environmental concerns. They licensed the technology from the holders of the old UK patent. Nishi had to overcome a massive setback just before beginning large-scale production because the heat-treated carbon for the anode proved less active and difficult to bind together than expected. Sony introduced its Li-ion battery in 1991 for use in camcorders and power tools, then new mobile phones.

Nishi said: *'At first, we assumed that important applications of LIBs were to audio and visual equipment such as cassette players, mini-disc players, home video cameras... two or three years after the first introduction of LIB we proposed using them for personal computer and cell phone manufactures – this brought about big changes to market.'*

Early LIBs earned a reputation for catching fire in a very dramatic fashion as they contained a formulation of lithium cobalt oxide that was prone to thermal runaway if the battery was ever accidentally overcharged. In 1996 a new formula for mixing LIBs was developed – lithium iron phosphate [$LiFePO_4$] or LFP cells, which are intrinsically non-combustible and so proved vastly safer than the early cells.

The 2019 Nobel Prize in Chemistry was awarded jointly to John Goodenough at the University of Texas at Austin, for making possible the development of LIBs and battery science, the chemist Stanley Whittingham of New York and Akira Yoshino of Tokyo, who each had made essential contributions to the development of these batteries.

High energy Lithium Manganese Dioxide 3v battery (2mm thick)

There were many attractions of the new lithium-ion rechargeable battery. They are lightweight, powerful, flexible and have long lives. A LIB is physically twice as small, three times lighter and capable of 10 times the number of recharge cycles than the equivalent lead-acid battery. Lithium batteries charge at nearly 100% efficiency, compared with the 85% efficiency of lead-acid batteries, which is important when charging from solar cells as it means most of the collected sun's energy goes into the batteries. With LIBs it is practical to use over 90% of their rated capacity, while the discharge curve of the batteries is essentially flat, so a 20% charged battery provides nearly the same output voltage as a fully charged battery. This makes LIBs well suited to powering high current loads like air conditioners, microwaves or induction cookers.

LIBs do not need to be stored upright or in a vented battery compartment. They can be manufactured and assembled into a variety of shapes and sizes. Manufacturers have found that tens of thousands of cycles can be run and they can be rapid charged to 100% of capacity at very fast rates. Unlike with lead-acid, a failure to fully charge LIBs regularly does not damage the batteries. Finally, they are much more efficient at low temperatures which have less impact on discharge rate and performance. This makes them suitable for use in extreme environments.

There are downsides for LIB technology given that lithium reserves are finite and it has environmental costs. Extracting the raw materials, mainly lithium and cobalt, requires large quantities of energy and water. Moreover, the mining often takes place in unsafe conditions or conflict areas.

Storage for renewables

LIBs are now used everywhere, commonly for example, to power portable electronic devices, such as mobile phones and laptops, or electric vehicles (EVs). They have revolutionized portable electronics and laid the foundations for the 'wireless' world. The batteries have helped transform the way people communicate, work, study, seek entertainment and search for knowledge. They now facilitate a renewables-powered world storing energy from solar and wind, which are often intermittent.

The cost of LIBs has fallen by 90% since 2010. Global sales to satisfy Li-ion battery demand for EVs and power storage alone recently amounted to about 60 GWh in a market worth over $50 billion. By 2040 sales of these batteries could increase to over 4,000 GWh.

In the future, a succession of improvements will continue to increase the energy density of LIBs to over 600 Wh capacity per litre of battery volume. Lithium-sulfur technology could allow cheaper batteries and offer a longer range for EVs. The possible replacement of graphite with silicon or tin will increase the energy density even further. Improvements to the cathodes would also boost energy density. The University of Illinois at Chicago recently developed a prototype for a novel design of a 'lithium-air' battery, which can store five times more energy than conventional Li-ion batteries.

M: Medicinal chemistry

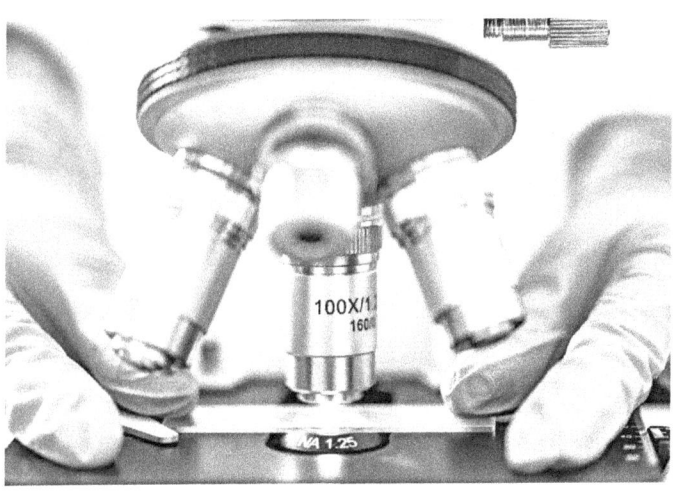

Finding 'magic bullets'

Medicinal chemistry involves the design and development of chemical compounds for use as medicines, for which chemical technologists and engineers then build and commission larger-scale facilities. Over the past 100 years a vast number of new medicines have had a major impact on health outcomes, overall quality of life and economic wellbeing.

Medicinal chemists first synthesize potentially useful compounds for biological evaluation that may prove effective against selected disease targets. They evaluate molecular structures to determine the chemical group responsible for evoking target biological effects. These structures can be changed, for example by inserting new chemical groups, to test their effects, efficacy and of course safety. Medicinal chemists engaged in drug discovery use their extensive knowledge of organic chemistry with data obtained from previous discoveries, laboratory research and medical professionals.

The roots of modern medicinal chemistry lie in the work of Paul Ehrlich in Germany, who envisaged a 'magic bullet' to combat infectious diseases. Medicinal chemistry now plays a major role in drug research and development, taking advantage of new technologies and knowledge drawn from biological and chemical sciences.

History

The idea of using chemicals to treat ailments can be traced back to ancient times, when it was noticed that certain plants had therapeutic effects. The ancient Chinese herb *ma huang* (*Ephedra sinica*) has been used to treat fevers since 3000 BC. The Sumerians knew of the narcotic effects of the opium poppy. In Ancient Egypt, a medical text known as the *Ebers Papyrus* contained numerous recipes and prescriptions using minerals and plants. The Egyptians believed that disease was caused by an imbalance of bodily fluids (*humors*), such as blood or bile, and that the balance could be restored using specific chemical treatments.

Similarly in Mesopotamia, minerals, metals and plant extracts were used for therapeutic purposes. In Ancient Greece, Hippocrates (460–370 BC) and Galen (129–216 AD) profoundly influenced experimentation to create medicinal remedies. Apple cider vinegar and honey were prescribed for coughs by physicians. The study of alchemy in the Medieval Islamic world further contributed to the advancement of medical chemistry by developing the techniques for distillation and extraction of active components from medicinal plants. Apothecaries used the opium poppy as an anaesthetic during amputations. In the Renaissance period, the early alchemists sought to discover the essence or active ingredient used in chemical medicines, especially the inorganic components such as mercury and antimony. The herbal works of John Gerard (1596), John Parkinson (1640) and Nicolas Culpeper (1649) provide an insight into the widespread use of herbs in medicine at the time. Increased global exploration from the 17th century led to tropical plants being used in addition to local metals, natural salts and earth.

The 19th century saw the beginnings of modern organic chemistry, particularly the isolation of several alkaloids from plants extracts including morphine (1805), quinine (1823) and atropine (1834) as part of the analytical effort to standardize drug preparations, control use and prevent fraud. General anaesthetics were introduced in surgery such as diethyl ether (1842), nitrous oxide (1845) and chloroform (1847), while antiseptics such as iodine (1839) and phenol (1869) made a vital contribution to the success of surgery. Joseph Lister in Glasgow was first to use a spray of phenol to disinfect wounds during surgery.

Some natural compounds were synthesized specifically for their medicinal action, such as the active constituent of willow bark (salicin) often employed as a painkiller in the 1860s. The acetylation of salicylic acid helped reduce its deleterious effect on the stomach. This led to the introduction of aspirin in 1899. Paracetamol (1878) was created as a painkiller for treating fevers. The local anaesthetic action of cocaine was reported in 1884, with many of its synthetic derivatives becoming medicines.

The first theories of the relationship between chemical structure and biological activity began to emerge. Edinburgh chemists Alexander Crum-Brown and Thomas Fraser noted in 1869 that *'a relationship exists between the physiological action of a substance and its chemical composition'*, leading to the idea that cells can respond to the signals from specific molecules. In the 1890s, after observing that certain dyes selectively stained micro-organisms, biochemist Paul Ehrlich put forward the idea that there were specific receptors for biologically active compounds – so-called 'lock and key' relationships. Following on from Pasteur's landmark work (see V and Y) it can be said that the field of medicinal chemistry was founded by Ehrlich, who is also regarded as the 'father of chemotherapy' for his extensive work on developing drugs to treat specific diseases.

Pioneer

Paul Ehrlich (1854–1915) developed a chemical theory to explain the body's immune response and suggested chemists would soon be able to produce substances that would seek out specific disease-causing agents.

Ehrlich was born near Breslau, then in Germany and now in Poland. He studied to become a medical doctor at the university and worked in the laboratory of his cousin Carl Weigert, a pathologist who pioneered the use of aniline dyes as biological stains. Ehrlich became interested in the selectivity of dyes for specific organs, tissues and cells, and he continued his investigations at a hospital in Berlin. After he showed that dyes react specifically with various components of blood cells and the cells of other tissues, he began to test dyes for therapeutic properties to find those that could kill off disease-causing microbes. He met with promising results using methylene blue, which attacked the malaria parasite.

After suffering from tuberculosis himself and receiving a cure from a therapy that was developed by bacteriologist Robert Koch (1843–1910), Ehrlich focused his attention on bacterial toxins and antitoxins. Ehrlich went to work at Koch's institute, which had recently developed 'serum therapies' for diphtheria and tetanus. Whereas Louis Pasteur's vaccines and Koch's tuberculin were made from weakened bacteria, these new serum therapies used blood serum, or cell-free blood liquid, extracted from the blood of naturally or artificially immunized animals to induce immunity.

Ehrlich developed his chemical theory to explain the formation of antitoxins, or antibodies, to fight the toxins released by the bacteria; at the same time Russian Elie Metchnikoff was studying the role of white blood corpuscles in destroying the bacteria. Most scientists agreed that both these explanations of the immune system were valid. In 1908 Ehrlich shared the Nobel Prize for medicine with Metchnikoff for their separate work to understand the immune response.

Ehrlich supposed that living cells have 'side chains' – a shorter chain or group of atoms attached to a principal chain in a molecule – much in the way that dye molecules were known to have side chains that were related to their colour. These side chains can link with particular toxins. According to Ehrlich, a cell under threat from foreign bodies grows more side chains, more than are necessary to lock in foreign bodies in its immediate vicinity. These 'extra' side chains break off to become antibodies and circulate throughout the body. It was these antibodies, in search of toxins, that Ehrlich first described as 'magic bullets'.

Infectious or communicative diseases were always a serious concern, but their causes were unknown and so there were no effective treatments. In Europe around 1880, 70% of cases were fatal, while today less than 0.5% of infectious diseases are fatal. Serum therapy was for Ehrlich the ideal method of combating infectious diseases. In cases for which effective serums could not be discovered, Ehrlich synthesized new chemicals, informed by his theory that the effectiveness of a therapeutic agent depended on its side chains. The modern concept of 'drug receptors' originated from Ehrlich's theory, which was advanced by Cambridge physiologist John Newport Langley, who went on to describe the concept of 'receptive' substances.

In Frankfurt, Ehrlich turned from his work on serum therapy to chemotherapies. He first targeted the protozoa that were known to be responsible for certain diseases, such as sleeping sickness, finding trypan red dye as a highly effective cure for that disease. In 1906 a research institute for chemotherapy was established under Ehrlich's direction. His institute gave the Hoechst and Cassella chemical companies the right to patent, manufacture and market preparations discovered at the institute. Researchers, now including organic chemists and bacteriologists, broadened the targeted micro-organisms to include a bacterium that had recently been identified as the cause of deadly syphilis. Beginning with an arsenic compound, and after extensive testing, they made an important breakthrough in 1909, discovering *Salvarsan*, the first modern antimicrobial drug for the treatment of syphilis.

In World War I, acriflavine and proflavine dyestuffs were used for the treatment of sepsis in wounds. The work of

Hans Horst Meyer and Charles Ernest Overton on anaesthesia to relate a physical property to biological activity were first attempts at statistical and quantitative measurement to optimize doses. All these pioneers had laid the foundation for the field of medicinal chemistry, leading to its rapid growth during the first half of the 20th century.

The 1920s saw the recognition of vitamin deficiency diseases, followed by treatments for tropical diseases such as dysentery and sleeping sickness, with the creation of synthetic antimalarials pamaquine (1926) and mepacrine (1932) to replace natural quinine. Researchers were able to define a group of molecules (bioisosteres) that are structurally similar, with similar physical or chemical properties. They produce broadly similar biological properties that can be used to modify the activity of the compound, reduce its toxicity, or alter the metabolism of the drug.

In 1935 Paul Domagk observed the antibacterial action of the sulfonamide dyestuff prontosil red, which led to development of an important family of sulfonamide antibacterials (*sulfa drugs*). Deaths from scarlet fever in Britain were soon on the decline. With the onset of World War II, there was an urgent need for new antibiotics in the military. Alexander Fleming had first observed that a strain of *Penicillium notatum* inhibited the growth of staphylococcus bacteria in 1928 (see Y). When in the early 1940s the β-lactam structure of penicillin was determined, a viable manufacturing process developed and a range of new penicillin antibiotics was made available with dramatic effect.

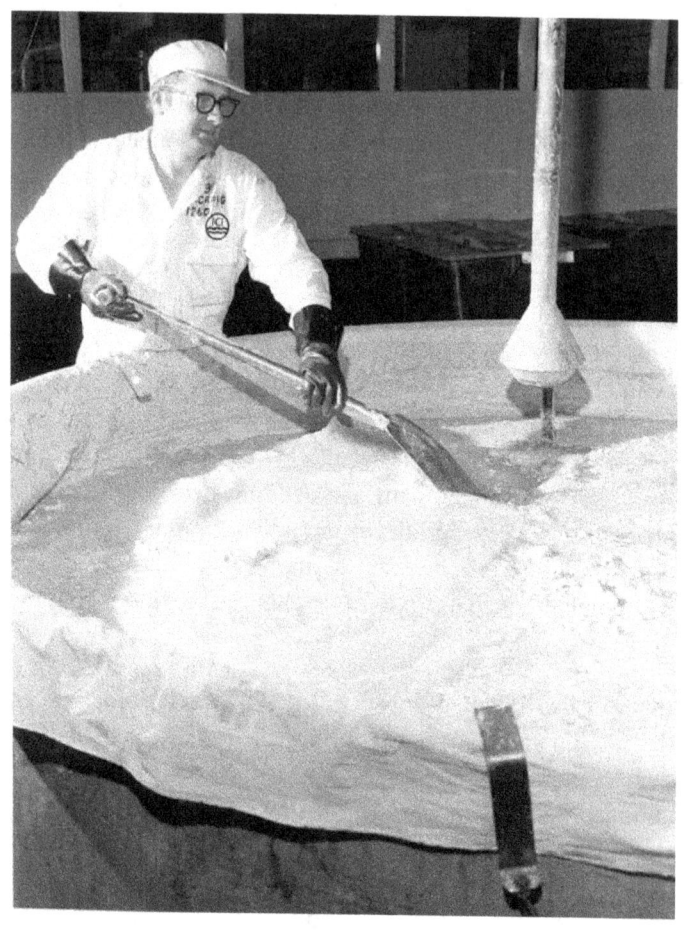

Washing and isolation of bulk Sulpha drugs at ICI (1974)

Drug discovery

Following the progress made during the war, drug companies began to organize to improve new drug discovery as advances in the understanding of pharmacology at the molecular level made it possible to determine the crucial structural features of a molecule that contribute to its biological activity. This later developed into structure activity relationship (SAR) studies, which are still a mainstream technique used in the drug discovery process. In 1962 the use of quantitative SAR studies in drug design was started by chemist Corwin Hansch (1918–2011) in California. He is known as the 'father of computer-assisted molecule drug design'.

In the 1960s several problems were to lead to safer medicines and more robust drug registration requirements. There was the serious case of the drug thalidomide, which had been introduced as a safe sedative but without adequate testing had been prescribed to pregnant women to treat morning sickness, causing many children to be born with deformities. The discovery of the beneficial effect of cortisone in alleviating the inflammation from rheumatism stimulated drug research, resulted in the development of several anti-inflammatory, semi-synthetic corticosteroids. In the second half of the 20th century came many breakthrough drugs, among which were chlorpromazine for the treatment of psychotic disorders (1953), propranolol, a beta blocker for high blood pressure (1965), beta2 agonists for the treatment of bronchial asthma (1969), tamoxifen for breast cancer (1969), cyclosporin, an immunosuppressant (1983) and antiretrovirals for HIV/AIDS (1987).

The long-term use of aspirin as a painkiller had brought side effects such as stomach ulcers, which required the development of non-steroidal anti-inflammatory agents (NSAID). The new painkilling drugs indomethacin and ibuprofen were introduced (1965 and 1971). The 1960s saw histamine antagonists for the treatment of peptic ulcers including cimetidine (1976) and then ranitidine (1981). Research to inhibit gastric acid secretion for patients with ulcers led to understanding the proton pump in the digestive system, resulting in the launch of lansoprazole (1989), the first proton-pump inhibitor (PPI), for the treatment of gastric acidity and ulcers.

Sometimes complex human biology can make targeting a specific molecule rather difficult, resulting in undesirable off-target effects. Occasionally these effects can be beneficial, as in the case of sildenafil, which was originally developed for blood pressure control but was later exploited for patients with erectile dysfunction or semaglutide, first used to treat type 2 diabetes but now also as an anti-obesity medication.

Drug discovery requires a mixture of chemical and biological knowledge, experimental findings, creativity and serendipity to identify potential target drugs. The process begins with the identification of a medical need and information on the adequacy of any existing therapies. From this and knowledge about the target disease many potential compounds for treatment are generated. Early on in drug discovery medicinal chemists relied primarily on data from the effects of chemicals tested on living organisms, rather than in test tubes. Nowadays drug development is much too expensive to be guided by trial-and-error testing, although identifying novel leads remains an extremely challenging task in terms of time and cost. Medicinal chemists also apply a multitude of new safety requirements when selecting promising test candidates.

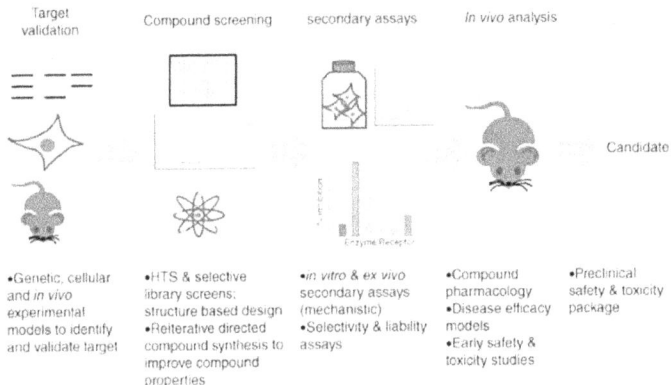

Target validation	Compound screening	secondary assays	*In vivo* analysis	
•Genetic, cellular and *in vivo* experimental models to identify and validate target	•HTS & selective library screens: structure based design •Reiterate directed compound synthesis to improve compound properties	•*in vitro* & *ex vivo* secondary assays (mechanistic) •Selectivity & liability assays	•Compound pharmacology •Disease efficacy models •Early safety & toxicity studies	•Preclinical safety & toxicity package

Process used to discover suitable new drug candidates

Testing of selected chemicals in appropriate biological situations was the next advancement. In this process, medicinal chemists try to detect relevant biological activity for a structurally novel compound in the laboratory (*in vitro*), then find a related compound with activity in an appropriate animal model (*in vivo*). This is followed by optimizing the activity through the preparation of analogous structures and finally selecting one compound as the drug-development candidate. This drug candidate then undergoes toxicological testing. If the compound passes all these tests, all the accumulated research data are assembled and submitted to the regulatory authority before clinical trials are initiated. In the clinic, there are a number of phases of evaluation, first in normal human volunteers to assess toleration (Phase I), then the efficacy and dose range in patients (Phase II), followed by widespread trials in thousands of appropriate patients to develop a broad database of efficacy and safety (Phase III). Only a very few drug candidates survive this series of trials before an application can be made for a licence to market it as a drug. All the accumulated research data is filed for review by the experts at the appropriate regulator. After formal regulatory approval the new drug can be manufactured and then launched commercially to doctors for prescribing to their patients to treat the disease for which it was designed. After launch any long-term effects of the drug must be monitored in a much larger population of patients (Phase IV).

It can take up to 10 years and over $2 billion to develop one successful major drug. Despite these significant investments in time and resources, some 90% of drug candidates in clinical trials fail because they do not adequately treat the condition, or the side effects are too serious or the drug proves uneconomic. This means many drug candidates never advance to the approval stage.

The role of the medicinal chemist in drug discovery has benefited more recently from the introduction of technologies such as combinatorial chemistry and computer-aided, structure-based drug design. The use of high-throughput screens – which combine automated robotics and micro-scale reactors to achieve the screening of hundreds of thousands of compounds with extremely small amounts of sample – has improved the efficiency of the drug-screening process.

The repertoire of candidates is no longer limited to small molecules but has expanded to biologics that include protein therapy, chemical probes and antibody–drug conjugates. Chemists still need sound knowledge of functional group chemistry of drug molecules, human physiology, drug administration and dosing. The increasing advancement of analytical techniques has allowed an understanding of how medicines are absorbed, distributed and metabolized by the human body.

A leading emerging technology is 'flow' chemistry. Compared with conventional batch processing, which is often carried out in flasks batchwise, continuous flow-based systems have demonstrated their potential in the rapid assembly of compound collections and in optimization and scale-up of drugs, and they are being applied to the manufacturing of active pharmaceutical ingredients (API). Flow chemistry allows for precise control over reaction conditions and enables real-time monitoring of reaction kinetics, resulting in higher quality products and cheaper, streamlined processes.

Personalized medicine

The foundations for medicinal chemistry were laid in the late 19th century when the early pharmaceutical industry saw that a more scientific approach to finding new cures and drugs would be necessary. Modern medicinal chemistry linked to chemical and pharmaceutical manufacturing technology has helped produce the new drugs that ultimately deliver many benefits and a higher quality of life to patients worldwide. It provides hope for people facing serious chronic illnesses and diseases while making traditional surgical treatments much safer by preventing infection or relieving pain.

In the future more 'personalized' medicine, which can provide treatments tailored to each patient based on their distinctive genetics, will come to the forefront of drug development. New techniques supported by informatics will assist in the rapid identification of novel drugs, reducing the costs of screening and clinical work to allow more rapid progress in the fight against cancers and genetic and infectious diseases.

N: Nitrogen fertilizers

Granular fertiliser providing critical nutrients to crops

Technology for feeding the world

Nitrogen fertilizers are essential to producing sufficient food for the world's population. They are produced from ammonia, which is now one of the most important chemicals globally, with approximately 85% being used in food production and the remainder in other nitrogen compounds, including explosives, pharmaceuticals, polymers, dyes and refrigerants. In the early 20th century, to meet increasing demands for food from agriculture, it was necessary to find a way to produce nitrogen fertilizers by converting atmospheric nitrogen to more reactive forms. This was made possible in 1913 by the Haber–Bosch process, which produced ammonia from hydrogen and air.

Nitrogen is one of the most important elements for life, being an essential component of both amino acids (the building blocks of proteins) and of nucleotides (the building blocks of DNA and RNA). It is also present in chlorophyll, the green pigment needed for photosynthesis by plants. While nitrogen makes up nearly four-fifths of the atmosphere in the form of nitrogen molecules [N_2], it is highly unreactive and not readily available for use by living organisms. Many commercially important crops, such as corn, rice or wheat, do not 'fix' nitrogen from air but absorb it from the soil and so it is necessary to add fertilizers in order replenish soils used for farming.

The groundbreaking Haber–Bosch process required specialized plants running at high pressures (around 200 bar) and high temperatures (at least 400 °C) and new catalysts. Industrially, air is the nitrogen source and natural gas (methane) often the hydrogen source. Ammonia production grew rapidly and made possible the intensification of nitrogen fertilizers globally, revolutionizing farming and increasing food production. It supported the expansion of food for a human population of around 1.5 billion in the early 20th century to 8 billion people by 2022. It is estimated that about half of the crops grown today are dependent on the nitrogen originating from ammonia-based fertilizers. It is said that the discovery and development of this chemical technology was of greater fundamental importance to the modern world than the airplane, nuclear energy, spaceflight or television.

History

In nature, breaking up nitrogen molecules requires the high energy from lightning or specialized microbes that are found in soil or live symbiotically in nodules of the roots of certain plants, such as legumes. The microbes use enzymes to convert nitrogen from the environment into the forms that plants can use as nutrients in a process called *fixation*, which turns nitrogen into organic nitrogen – the form combined with carbon in a wide variety of molecules essential to both plants and the animals that eat them. By the process of denitrification, organisms use nitrogen nutrients as their energy source then return nitrogen molecules to the atmosphere, completing the nitrogen cycle.

Before synthetic fertilizers were available, farmers made use of seaweed, animal bones and manure to improve soil fertility. In 1802 German explorer Alexander von Humboldt collected samples of bird guano (sea-bird excrement) in Peru when told that indigenous peoples used it to greatly improve the growth of plants, being an effective fertilizer due to its exceptionally high nitrogen, phosphate and potassium content. In 1813 the famous British chemist Sir Humphry Davy promoted the benefits to farmers of using a balanced chemical fertilizer from crushed animal bones. Soon bone-grinding mills were in operation producing fertilizers for applying to fields. In 1841 John Bennett Lawes (1814–1900), an entrepreneur and agricultural scientist, investigated the nutritional needs of plants and discovered that elements such as nitrogen, potassium and phosphorous were essential. He began solubilizing bones with sulfuric to make superphosphates at his Deptford factory. This was the start of the artificial fertilizer industry.

Then large guano deposits (45 m deep) were found to have covered the surface of the Chincha Islands, off the coast of

Peru, allowing over 200,000 tonnes a year of guano to be shipped overseas. Though the guano trade created a boom in the Peruvian economy, it also encouraged international rivalry, as did the emergence of a new fertilizer industry based on the mineral caliche (containing sodium nitrate), which was found in the desolate Atacama Desert of Chile. Disputes over rights to the Atacama deposits led to a war in 1879. The Chilean nitrate industry became the main supplier of nitrogen for agriculture and the explosives industries. This monopoly caused the US to take possession of many unoccupied guano islands, while Europeans with limited saltpetre (sodium nitrate) reserves urgently explored the chemical means to 'fix' atmospheric nitrogen, as signs of an impending food production crisis emerged.

In 1898 the British chemist Sir William Crookes foresaw the crisis arising from the massive growth in population, with the advent of the industrial age and migration to cities but only limited supplies of natural fertilizers for improving soil nutrients – a solution was needed or famine was certain. A new chemical technology to increase fertilizer production was urgently required to achieve the fixation of atmospheric nitrogen. Nitrogen gas, although abundant in the atmosphere, has triple bonds that make the molecule incredibly stable and hard to react.

The fundamental process to make ammonia involves the reversible reaction between nitrogen and hydrogen: $N_2(g) + 3H_2(g) \rightleftharpoons 2NH_3(g)$. Worldwide population, food supplies and living standards would rise dramatically once there were reliable and economic supplies of ammonia-based nitrogen fertilizers. The successful new chemical technology for the synthesis of ammonia from nitrogen and hydrogen was pioneered in Germany.

Pioneers

Fritz Haber (1868–1934) was one of a group of chemists who decided to tackle the problem. Haber at first experimented with producing nitric oxide from electric discharges, mimicking the natural processes in a thunderstorm, but the yield was very low and the process was dismissed as impractical. In 1905 he also investigated high-temperature synthesis at 1,000°C, but this was uneconomic giving only a 5% yield. He needed a catalyst and higher pressures, but high-pressure synthesis was then in its infancy. Indeed, Henry Louis Le Chatelier, the French chemist whose 'principle' was used by chemists and chemical engineers to predict the effect of changing conditions in a chemical equilibrium, was the first to suggest fixing nitrogen under high pressure, but he had given up his own experiments after a serious explosion in his laboratory. It was not until 1908 that Haber, working with his British student Robert Le Rossignol, decided to investigate a high-pressure route. After a year, they patented a new process operating at 175 atmospheres at 550°C employing an osmium and uranium catalyst. The process yielded about 15% ammonia. His work making ammonia from hydrogen and nitrogen gases was financed by BASF, the German dye and chemical manufacturing firm, which could undertake difficult and expensive scientific projects. BASF asked its chemical engineer Carl Bosch to scale up Haber's laboratory process. Bosch later said: *'It was obvious that there were three main problems to be settled before the construction of a plant could be undertaken. These were supply of raw materials, the gases hydrogen and nitrogen, at a lower price than hitherto possible; the manufacture of effective and stable catalysts; and lastly the construction of the apparatus.'*

Bosch obtained pure hydrogen by cooling 'water gas' containing hydrogen and carbon monoxide produced from coal-synthesis gas (to −205 °C), liquefying all apart from the hydrogen stream. The original osmium-uranium catalyst was scarce and very sensitive to water and oxygen. It took Bosch's assistant, chemist Alwin Mittasch, a pioneer of catalytic chemistry, thousands of experiments to perfect an alternative catalyst using iron oxide.

Bosch thought his greatest achievement was designing and building a reactor that would be robust and reliable enough to withstand both the high temperatures and high pressures of the reaction. High-pressure chemistry was a new field, and suitable equipment was not available. Bosch first built a new laboratory reactor with small reaction chambers. The outer pressure-bearing parts were small enough that air cooling was sufficient to keep them stable, and there was less mechanical stress on the inner parts. Haber was not able to scale up this reactor as it failed after only a couple of hours of operation, so Bosch's team designed a very sturdy, externally heated contact tube as a reaction chamber. He said he placed it in a strong, reinforced concrete chamber far away from the other buildings because of the *'danger of fires and flarebacks which occur, frequently with spontaneous ignition, when hydrogen emerges at high pressure'*. It was a wise precaution: after 80 hours of service, the metal became brittle and the tubes burst.

Bosch innovatively solved the problem by designing the first lined reaction chamber – a pressure-bearing steel jacket thinly lined with a soft steel. Hydrogen was able to diffuse through the lining and was allowed to escape through holes in the jacket, to prevent a dangerous pressure build-up. Always very conscious of safety, he designed numerous quick-acting safety valves and other equipment so that the plant could be shut down and evacuated rapidly. He found a means of heating the reactor from the inside, building new high-pressure gas compressors and creating monitoring instruments to measure temperature, the flow of the gas stream and the composition of the gas in the reaction chamber. Bosch is considered one of the great pioneers of chemical engineering.

In 1913 the Haber–Bosch process produced its first ammonia on an industrial scale at BASF's works at Oppau. When World War I began, the British Navy had blocked Germany's access to Chilean nitrates for explosives, but the country was able to convert its synthetic ammonia from the new Haber–Bosch process into explosives. After the war it was more economic to use the new Haber–Bosch process rather than to import Chilean nitrate. This process was soon adopted around the world. Ammonia synthesis by the Haber–Bosch process was the only chemical breakthrough recognized by two Nobel Prizes for Chemistry, awarded to Fritz Haber in 1918 and then again to Carl Bosch in 1931. Bosch went on to be a founder of the large German company IG Farben.

Fritz Haber in his laboratory (L) Carl Bosch chemical engineer (R)

Germany used the synthetic ammonia from the Haber–Bosch process to pave the way for inexpensive fertilizers, providing enormous benefits to agriculture and global development. Haber's reputation was later harmed by his work to advance large-scale use of chemical weapons. Such weapons resulted in 100,000 fatalities and the Allies deemed Haber a war criminal. He was forced into exile in Switzerland, to die a broken man in 1934. In 1997 the Chemical Weapons Convention came into force banning the development, production, stockpiling and use of all chemical weapons worldwide.

In Britain the first commercial ammonia plant was established at Billingham in north-east England in the early 1920s. This again grew out of the needs of World War I because of the shortage of chemicals for munitions. There were insufficient nitrate-based explosives as nitrate supplies were suffering attacks by German naval submarines. The British government set up the factory to develop the technology for synthetic ammonia which after the war it was taken over by Brunner Mond (later part of ICI), and converted it to make ammonia-based fertilizers. The first Billingham plant initially produced only 24 tonnes per day of ammonia.

Through the 20th century, ammonia production technology continued to advance under the competitive pressure to increase output with more energy efficiency, to lower environmental emissions and increase plant reliability. The first ammonia plants built at Billingham employed the classic Haber–Bosch process based on coke. By the late 1950s, with output running at over 750 tonnes per day, the increasing cost of coal and the inefficiency of generating syngas from coal were making this process uncompetitive. In 1962 the syngas units at Billingham were modernized using four naphtha (light distillate hydrocarbon) steam-reforming units. In the 1970s Billingham's ammonia plants were again converted from the naphtha feedstock to run on natural gas supplied from the North Sea. As a result of improvements in metallurgy the reactor catalyst beds were able to operate at higher temperatures and pressures.

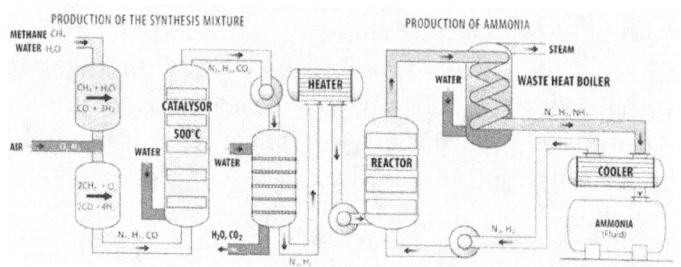

Flow diagram of the Haber-Bosch ammonia process

Modern production

Worldwide over 70% of ammonia production uses hydrogen obtained from the steam reforming of natural gas (methane) or other light hydrocarbons. Pressurized to 200 atmospheres, nitrogen (extracted from the air) and hydrogen are heated to 450 °C and passed through the main reactor containing an iron catalyst. The reaction mixture is then cooled to liquefy the ammonia for storage. Unreacted nitrogen and hydrogen gases are recycled. The production of ammonia is a very energy-demanding process, consuming 30 GJ per tonne of ammonia. In the old established plants, as the energy utilization was nearing the theoretical minimum there was little scope to reduce energy consumption. The ammonia plants in Billingham were finally closed in the 2020s due to the high costs of natural gas and energy taxes in the UK.

Globally, advances in catalyst and process technology continued into the 21st century. The British manufacturer ICI and German technology company Uhde developed the 'dual-pressure ammonia' process using several innovative concepts. Their first ammonia plant was the SAFCO IV plant in Al Jubail, Saudi Arabia, which was started up in 2006 with a capacity of 3,300 tonnes per day – one of the largest ammonia plants worldwide. The larger plants offer a significant reduction in production costs (see illustration of a modern ammonia plant on page 11).

The fertilizer industry manufactures many products such as ammonium nitrate, ammonium sulfate, urea, ammonium phosphates and mixtures of these fertilizers. Anhydrous ammonia is also used as the most concentrated nitrogen fertilizer. The ammonia liquid is injected under pressure into the subsoil, where it reverts to a gaseous state and combines with the soil moisture to provide the essential nitrogen nutrient. The global production of synthetic ammonia is over 175 million tonnes per year and continues to grow to satisfy the increasing demands for fertilizers for food and ammonia-based intermediates for the ever-increasing global population. As the demand continues to rise so the environmental, human health and climatic aspects of ammonia and fertilizers present an additional challenge.

Nitrogen fixed from the air by industrial activity can contribute to a number of environmental problems, such as the worsening of the greenhouse effect from fertilizers releasing nitrous oxide [N_2O], which is over 200 times more potent a greenhouse gas than carbon dioxide; nitrous oxide also depletes the protective ozone layer, where sunlight breaks it into nitric oxide free radicals (see U); and finally it causes a harmful smog in the air, producing aerosols that might induce serious respiratory illness, cancer and cardiac disease. Less than half of the fixed nitrogen generated by farming practices ends up incorporated in harvested crops. New techniques are being employed for applying fertilizer more efficiently to reduce runoff, leaching and erosion, and to keep it away from groundwater or surface water where it over-enriches water courses. This leads to large amounts of phytoplankton (small water plants) that deplete oxygen in the water for other creatures.

Overall environmental policies require the shift from energy derived from fossil fuels to generation from renewable energy sources (solar, wind, tidal and geothermal). The Haber–Bosch process is one of the largest industrial global energy consumers and greenhouse gas emitters, responsible for 1.2% of the global anthropogenic carbon dioxide emissions, so that alternative processes are being investigated. In the future, plants will be using 'green' hydrogen produced without fossil fuels from the electrolysis of water with renewable energy.

Ammonia revolution

In just over 100 years, the nitrogen fertilizer industry based on ammonia production has grown massively. The ammonia synthesis process developed by Haber and Bosch that enabled large-scale production transformed food production for the growing global population. The core process can still be clearly recognized in the most modern ammonia plants, although the process efficiency is vastly better. The post-war years saw a profound revolution in agriculture, with synthetic fertilizers and mechanization resulting in a much-reduced need for labour and enormously increased crop outputs. In more recent decades, food production has been able to keep pace with human population growth thanks to the development of new high-yielding crop varieties optimally grown and increased use of synthetic fertilizers.

The global challenge to managing high energy demand is now driving interest in secondary 'energy carriers' like hydrogen from renewable energy and potentially ammonia. This could lead to a second ammonia revolution, helping to store energy to balance the seasonal energy demands with carbon-free renewables. Ammonia is a potentially good energy carrier as it provides an attractive alternative to the short-term storage offered by electrochemical storage in batteries. As a fuel ammonia has greater energy density than liquid hydrogen and an existing infrastructure that serves the fertilizer market – it will help the world achieve net zero carbon emissions, making it one of the most effective ways to transport and store hydrogen. This means 'green' energy from sunny and windy regions or from hydropower can be transported over long distances using ammonia, which is subsequently reconverted to hydrogen or used as a chemical feedstock.

O: Oral contraceptives

Family planning

The contraceptive or birth control pill was created by chemists as a type of chemical birth control taken orally by women. It contains synthetic versions of the hormones oestrogen and progesterone, which are produced naturally in the ovaries. When taken it alters the menstrual cycle to eliminate ovulation (release of eggs from the ovaries) and prevent pregnancy.

The oral contraceptive pill was introduced in the early 1960s and its use spread rapidly, leading to an enormous social impact. It was more effective than most previous methods of birth control, giving women new choices and unprecedented control over their fertility. This new biochemical technology with its ability to control fertility was influential in facilitating women's modern economic role and improving gender equality, postponing the age at which women started families and allowing them to remain in education and pursue careers. Effective family planning also meant there were far fewer unwanted pregnancies and a sharp drop in abortions or health problems due to constant pregnancies.

It was George Rosenkranz, a pioneering steroid researcher, who was instrumental in developing this world-changing pharmaceutical. In the early 1950s Rosenkranz was research director at a Mexican chemicals company. His team synthesized a substance they called *norethindrone* – the progesterone that was to be used in one of the first combined oral contraceptive pills and is still used by millions of women around the world.

Pioneer

George Rosenkranz (1916–2019) was born in Budapest, the only child of successful middle-class parents. His childhood gave him an appreciation for art, music, and theatre, and although he demonstrated a gift for modern languages, his scientific studies, particularly in chemistry, held a stronger appeal. In 1933 he attended the Swiss Federal Institute of Technology in Zurich, where he took an organic chemistry course taught by the future Nobel laureate Leopold Ruzicka. Rosenkranz was inspired by Ruzicka, who was working to synthesize the male sex hormone testosterone. He became a doctoral student and received his degree in 1940.

As World War II war progressed, being of Jewish decent, he was forced to leave Zurich and make a distressing escape from Europe. As a result of the attack on Pearl Harbour in 1941 he became stranded in Cuba, where he found a job in Vieta-Plasencia Laboratorios developing successful treatments for venereal disease. In 1945 Rosenkranz became head chemist in a company called Syntex in Mexico City. Syntex was a research firm founded just a year earlier that was attempting to make synthetic hormones. He was to spend his entire career at Syntex, eventually as its chief executive and chairman.

The idea of a contraceptive pill based on hormones had first been proposed in 1921 by the Austrian physiologist Ludwig Haberlandt. However, identifying, synthesizing and correctly dosing the right hormones was not straightforward – and then there were ethical challenges to overcome. Research on contraceptives was then illegal in many countries and medical scientists were reluctant to participate in this controversial area. It had become clear that oestrogen and progesterone hormones had a role in preventing ovulation. At the time, these had to be extracted from animal glands, a laborious process that made any hormone-based treatments extremely expensive. As the relevant patents were all held by European companies this gave them a competitive stranglehold over any US-based researchers.

Russell Marker, a maverick professor of organic chemistry at Pennsylvania State College, succeeded in revolutionizing the progesterone production process by synthesizing progesterone from sapogenins: the natural steroids found in Mexican yams. Marker developed a process for turning diogenin, a steroidal sapogenin, into progesterone by a five-step process, which brought the price down by two-thirds and destroyed the European cartel as the price fell from $80/g to less than $2/g. Syntex was formed in Mexico to commercialize the process. Within a year, Marker left the company, so it fell to its new employee George Rosenkranz to recreate Marker's process and restart the large-scale production of progesterone. Marker's laboratory records were inadequate, so Rosenkranz had to use his own knowledge of steroids to recreate the process. He succeeded and within two months the production of progesterone was back on track.

Rosenkranz was joined by Austrian Carl Djerassi and Luis Miramontes from the National Autonomous University of Mexico, working on his chemical engineering PhD. They set out to synthesize a progesterone steroid with the right characteristics. Miramontes did most of the practical work, using toluene as a solvent instead of benzene, as the latter's low boiling point made it unsuitable for the high altitude of Mexico City. On 15th October 1951 Miramontes completed the synthesis of the hormone *norethindrone* (19-nor-17α-ethynyltestosterone) – the most active oral pregestational hormone of its time.

The success of this work led to investment into Syntex and the Mexican steroid pharmaceutical industry. At that time it was not used as a contraceptive but to treat certain menstrual disorders and infertility. It was then realized that progesterone, especially when combined with oestrogen, was an effective contraceptive. Working on contraceptives remined problematic as scientists feared boycotts due to potential social and religious objections.

The contraceptive pill was eventually developed by American scientists led by Gregory Pincus at the Worcester Foundation for Experimental Biology in Massachusetts and with the active support of women's rights campaigners Margaret Sanger and Katharine Dexter McCormick. Hormone-based contraceptives mimic the conditions of pregnancy and therefore offset conception. A progesterone pill works by thickening the mucus in the cervix to stop the sperm reaching the egg.

A range of steroids were tested for efficacy, then two candidates were selected: Syntex's *norethindrone* and an isomer *norethynodrel*, synthesized in 1960 by Frank Corlton at the US drug manufacturer G.D. Searle. Ultimately, both would make it to market as the first combined oral contraceptive pill. The pill was first licensed for contraceptive use by the US Food and Drug Administration in June 1960. In the UK, following clinical trials, the Ministry of Health approved its availability on the NHS. As oral contraception required medical advice to prescribe it, the medical profession became more involved in providing family planning services and its use rapidly increased. The first commercially available oral contraceptive was branded *Enovid* in the US and *Enavid* in Britain.

George Rosenkranz and Carl Djerassi went on to gain further international recognition for Syntex, making it the first company to develop an industrially viable synthesis process for cortisone, another steroid that attracted huge interest as a treatment for rheumatoid arthritis. Until 1951 the only source of cortisone was through an extraordinarily complex process having 36 different chemical transformations starting from animal bile acids. It was considered for many years the longest and most complicated synthesis of any chemical on an industrial scale. The race for more economic cortisone captured much interest, with Syntex being up against many much larger and better-funded rivals. Although one of its rivals, Upjohn, would make the process obsolete by using a novel biochemical fermentation that shortened manufacture to a single step; fortunately this new process relied on Syntex's progesterone as a raw material, so the company still made money.

Over the years Rosenkranz and his colleagues created several revolutionary advances in the understanding of and production techniques for steroid drugs, often using native Mexican plant sources as raw materials. Rosenkranz was responsible for over 150 patents and he published numerous research articles on steroid hormones. He was a member of many learned institutions and several companies. He died age 102 in 2019.

Mexicans are proud that it was their researchers who were the first to synthesize an oral contraceptive pill, which contributed to social advance and allowed medical improvements all over the world. Eventually Syntex moved its research base to Palo Alto in California but retained its manufacturing plant for bulk steroid intermediates in Mexico, while producing finished drugs in Puerto Rico and the Bahamas. Syntex was acquired by the Swiss Roche group in 1994.

Making choices

Introduction of the synthetic birth control pill in the early 1960s had an impact on many aspects of women's health, fertility trends and government policies, sexual relationships and family roles, and women's rights. This allowed women to make new choices about their lives.

The oral pill proved to be highly effective from the outset. Although some safety issues developed with the earlier formulations, continued development has resulted in greatly safer oral contraceptives. Over the past 40 years, both the content and doses of the steroid components of the pill have changed significantly to avoid adverse effects. Medical consultation is advised if any side effects are experienced. More recently, the emphasis has shifted from the health risks of use to the non-contraceptive health benefits. By the end of their reproductive years, more than 80% of Western women will have used oral contraceptives, typically for an average of 5 years. On the other hand, progress in developing new reversible male contraception has been slow and there remains no commercially available product.

The UN believes that it is still the case that over 200 million women in developing countries have an unmet need for modern contraception. The pill can be seen as a force for sustainability facilitated by chemical technology. There has been a sharp decline in the birth rate in countries where the pill is available with family planning impacting women's health and having a positive economic effect on families.

P: Pesticides

Aerial dusting vegetable seedlings with insecticide (1940)

'Miracle' with a sting in its tail

Pesticides are substances intended for preventing, controlling or destroying pests. Such pests may be the cause of plant, human or animal disease, unwanted weeds or rodents causing harm to the production or storage of foodstuffs. There are naturally occurring or synthetic pesticides as well as biological agents such as viruses or bacteria.

It was chemical technology that produced the first very effective synthetic insecticide – DDT – used with great effect to combat malaria, typhus and other insect-borne human diseases and to control insects in crop and livestock production, institutions, homes and gardens. Through time and with greater understanding, it was seen that the indiscriminate application of pesticides could pollute water, damage bird, animal and native plant populations and in some cases cause severe medical problems for humans. In the modern era, pest management and control are based on a diverse science with many different strategies, sometimes using chemical products.

There are many common crop pests, such as those in wheat crops of leaf rust, fusarium head blight, aphids and powdery mildew; in rice, sheath blight, stem borers or brown spot; or in maize there is stalk rot, leaf blight and rust. These all cause significant losses globally. Staple crops such as potatoes suffer blights, scab, brown rot and attack from cyst nematode, while soybeans are affected by nematodes, mould and rust.

In addition to the many pathogens above, weed plants compete with farmers for food, as do rodents, which can spread disease or destroy property. Half of all emerging plant diseases are spread through global travel and trade. Climate change may increase the risk of pests spreading in agriculture and forestry, especially in regions experiencing unusually warm winters, which promotes the establishment of invasive pests and plant diseases. Desert locusts, one of the most destructive migratory insects, are expected to change their migratory routes and geographical distribution as the climate changes.

The damage from disease and plant pests leaves millions of people without enough food and reduces the yields and quality of agricultural production. Globally, farmers lose as much as 40% of their crops to pests and plant diseases each year at a cost of over $200 billion. This leads to substantial hunger, reduced food security and economic loss.

In the epic battle against pests and parasites starting from the early 20th century, chemical fungicides were developed for the control of fungi, herbicides for the control of weeds and insecticides for the control of insects. Soon householders, gardeners and horticulturists could buy a large selection of poisonous compounds, smoke capsules and liquid products, many for glasshouses, before the availability of synthetic chemical pesticides.

History

Prior to the development of synthetic pesticides there were limited means to reduce the onslaught of pests on crops. The early types of pesticides were botanical and natural substances. Crop dusting to deter insects with powdered sulfur was used by Sumerians some 4,500 years ago. A thousand years later in China, chalk and wood ash removed pests from stored grains. The Romans discovered that crushed olive pits could produce an oil that killed many pests. By the 15th century, toxic metal compounds, such as arsenic, mercury and lead, were being applied to crops. Nicotine sulfate was also extracted from tobacco leaves for use as an insecticide.

The chemical pesticide industry grew in the 19th century to employ many more inorganic compounds. Sulfur continued to be used for animal dips to destroy lice; sulfur dioxide was generated by burning elemental sulfur to inhibit the respiration of insects and other small pests. The toxic heavy metal compounds of arsenic and mercury were fatal to insects, bacteria and fungi because they interrupted biological or enzymatic processes. They were also used in wood treatment and preservative products. While long-lasting they easily degraded, leaching into the soil and watercourses, so posed several health threats to wildlife and people.

Many chemical compounds used for pest control were often borrowed from other industries. In 1814 the pigment 'Paris Green' (copper acetoarsenite) was widely sold as an insecticide and rodenticide. Similarly, copper salts were used to treat downy mildew in the vineyards of the French Bordeaux region, after it was noted that vines closest to the roads did not show mildew. These had been sprayed with a mixture of copper sulfate and calcium hydroxide to deter passers-by from eating the grapes by making them bitter-tasting. After field trials in 1885 the 'Bordeaux Mixture' was soon employed widely to fight serious fungal infection in vineyards.

As experimental investigations in chemistry and biology were increasingly undertaken so new chemical synthesis was to improve pest control. This led to identification of compounds that were historically used in their natural forms (such as nicotine from tobacco, cyanides in fruit seeds, rotenone from derris roots and pyrethrums in chrysanthemum flowers) being extracted and purified to be used industrially. However, supplies of natural insecticides were insufficient to meet the demands of farmers. Chemical technology joined the battle with the landmark discovery in 1939 by the Swiss chemist Paul Müller of the chemical compound commonly known as DDT. This first modern synthetic pesticide was an organochloride compound, one initially prepared by Othmar Ziedler, an Austrian chemist. Müller discovered that this compound had insecticidal properties. The compound appeared to be incredibly effective at eliminating the insects transmitting serious human and animal diseases. It was soon hailed as a 'miracle' insecticide as it became vital during World War II to control insects that were spreading typhus, malaria and dengue fever. It was employed for insect control in crop and livestock production, within buildings and in gardens. It became the most widely used pesticide in the world.

Pioneer

Paul Hermann Müller (1899–1965) was born in Switzerland. His father worked for the Swiss Federal Railway in Basle, where Paul attended secondary school. He commenced work in 1916 as a laboratory assistant at Dreyfus and Company, and the following year he joined Lonza A.G. as an assistant chemist, gaining a wealth of practical knowledge prior to studying for a doctorate at Basle University in 1925.

He became an industrial chemist at the great dye-manufacturing firm of J.R. Geigy, S.A. in Basel. Müller's first research concerned the electrochemical oxidation of *m*-xylidine, vegetable dyes and natural tanning agents. As a remarkably skilful and creative laboratory chemist, he developed several new synthetic leather-tanning products. He worked on disinfectants and moth-proofing agents for textiles, and in 1935 he developed *Graminone*, a seed treatment. He then began his research on new synthetic contact insecticides. Within a few years he had invented two new insecticides, trade-named *Gesarol* and *Neocide*; their specific toxic component, however, remained unknown to him.

In 1939, in search of their active ingredient, he synthesized a chlorinated hydrocarbon compound [1,1'-(2,2,2-

Trichloroethane-1,1-diyl) bis(4-chlorobenzene)], abbreviated as DDT. After tests, he found he had discovered the most powerful synthetic insecticide then known, as it proved fatal to an incredibly wide range of insects on contact in extremely minute quantities but was apparently not toxic to humans. Initially, DDT was seen as a broad-spectrum insecticide. It was inexpensive to produce, easy to apply to large areas and was very persistent; being insoluble in water it was not easily washed away by rain and had low toxicity to mammals. This potent new insecticide was granted a Swiss patent in 1940. Field trials showed it to be effective not only against the common housefly, but also against a wide variety of pests, including the louse, Colorado beetle and mosquitoes.

Paul Müller became deputy director of Scientific Research on Substances for Plant Protection in 1946. He was honoured throughout the world for his discovery, receiving a Nobel Prize in 1948 for Medicine for his positive impact on global health.

Swarm of damaging desert locusts in Kenya

DDT successes

The new products based on DDT soon came to the attention of British and American medical entomologists as wartime supplies of natural pyrethrum insecticide were not able to satisfy the high demands. The Allied army and navy urgently sought more insecticides to control malaria (carried by mosquitoes), epidemic typhus (carried by body lice), dysentery and typhoid fever (both carried by houseflies). These diseases could kill more people than were being lost in fighting. DDT provided a powerful synthetic contact insecticide that was safe, easy to handle and capable of being economically mass-produced. Millions of soldiers and sailors carried small cans of DDT powder to protect themselves against bedbugs, lice and mosquitoes in the tropics, while DDT aerosol sprayers were used to treat the interiors of tents, barracks and mess halls. Large-scale DDT production was soon established, which proved of enormous value in combating and eliminating typhus and malaria. By 1943 Geigy's production was approaching 1,500 tonnes a month. This provided a great stimulus in the search for other uses for the insecticide.

The insecticide was widely used in refugee camps to control infestations of stinging, biting and disease-spreading insects. The success of field trials in Egypt to control a small typhus epidemic led to its use in Italy, where Allied medical authorities saw that a major epidemic was progressing in the newly liberated, refugee-swollen city of Naples. Cases approached 60 a day, with people dying everywhere. If the epidemic had followed the usual exponential explosion, it would have led to 250,000 deaths. By December 1943, following the systematic dusting of the entire population of the city with DDT, the number of new cases per day was in sharp decline. By the next February there were none. For the first time in history, despite the filthy, overcrowded conditions that perfectly suited typhus, it was rapidly eliminated. US newspapers announced that, thanks to the new DDT pesticide, *'dreaded plague that has followed in the wake of every great war in history is no longer a threat to American and Allied troops.'*

In the southern US, DDT pesticide was used for killing malaria-carrying mosquitoes, protecting people and their animals. While the insecticide was clearly seen as poisonous, insects did not have to eat it but merely come into contact with it, and it remained effective for months after it was applied. DDT killed an extraordinary range of insects at very low doses without causing any detectable harm to people. By the early 1950s, annual DDT production had increased ten-fold to more than 45,000 tonnes. The majority was used in agriculture, where there was an urgent need to increase food supplies as over 30% of crops were being lost to insect damage.

Researchers at the US Department of Agriculture did start to notice some problems in the field. When a solution of DDT was sprayed over an oak forest in Pennsylvania infested with gypsy moth it was completely effective as it killed every caterpillar within hours; but after a week at least 4,000 birds had also died. Further, the spraying had resulted in a tremendous multiplication of aphids that defoliated trees, which previously were naturally controlled by ladybirds. The product was killing many beneficial and harmless insects that were needed as crop pollinators. The DDT was extremely persistent due to its chemical stability and insolubility in water. It remained fatal to flies and mosquitoes for up to three months, a treated mattress was kept free of bedbugs for nine months and a DDT-sprayed blanket could be laundered a half dozen times but still kill any moth that touched it. While this was of great advantage in both the military and in homes, some scientists questioned the need for mass spraying of fields and orchards, forests and pastures, city parks and tree-lined streets. Nevertheless, despite these early concerns, DDT was released for almost unrestricted agricultural, household and garden use in 1947.

There is no doubt that using DDT seemed overwhelmingly justified by its benefits. Within five years, food production increased, millions were saved from starvation and five million lives were saved globally due to the destruction of malarial mosquitoes. In Greece, where a third of the workforce had been losing several months of work time each year to malaria and where malarial infant mortality in many villages approached 100%, the disease was virtually eliminated by a massive DDT-spraying campaign. Müller's work had saved lives and reduced diseases such as typhus, malaria, yellow fever and plague.

By 1963 US production of DDT had reached its annual peak of over 80,000 tonnes. The World Health Organization stated the insecticide had prevented the death of 25 million people since World War II. But its success and its widespread use did begin to lead to problems: the development of resistance in many insect species, harm to many animals and threats to the environment. It was the most well-known synthetic chemical in the world, but only a few people continued to question its potential side effects. One who did was a pioneering young biologist, who about the time Müller had begun his pesticide research was teaching in the zoology department at the University of Maryland.

Unwelcome side effects

This biologist was Rachel Carson (1907–1964), who was born in Springdale, a rural town near Pittsburgh. In 1925 she entered the Pennsylvania College for Women to pursue a writing career, while also taking classes in biology. She graduated with honours, completing a master's degree in zoology from Johns Hopkins University, followed by postgraduate work at the Marine Biological Laboratory in Woods Hole, Massachusetts.

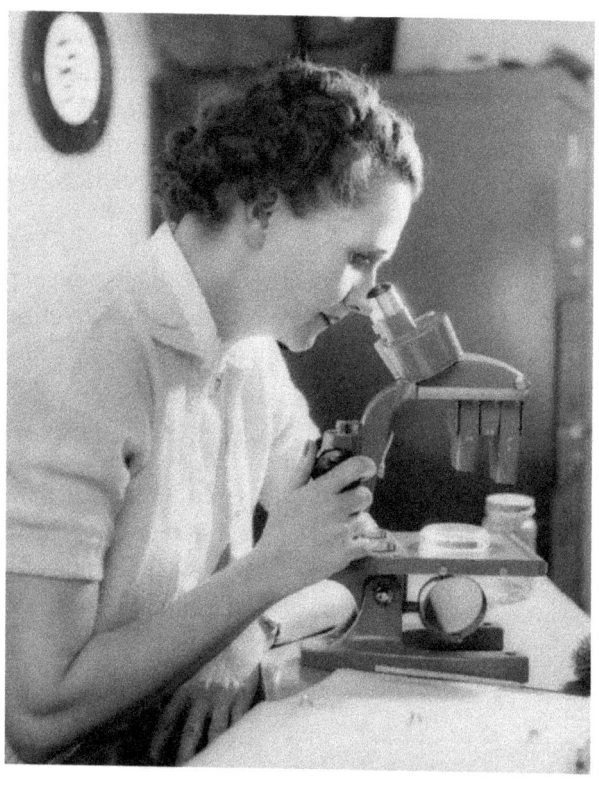

Rachel Carson marine biologist and conservationist

Carson's fascination with science coupled with her desire to protect natural areas inspired the rest of her life. She joined the US Bureau of Fisheries as an editor in the Fish and Wildlife Service. In 1951 her book *The Sea Around Us* secured her reputation for writing with scientific thoroughness but with an engaging style. The book was a bestseller, leading her to resign her government job to devote herself to research and writing on environmental matters. She had begun to read reports on the new 'miracle' insecticide that disturbed her. She heard that DDT-resistant strains of houseflies, mosquitoes and crop-destroying insects had begun to appear naturally, which needed massive doses of the insecticide to control.

She understood that DDT was directly poisoning the environment, including birds, fish and small game. In 1958 she saw a crop-spraying plane over a two-acre coastal bird sanctuary in Massachusetts, then next morning, visiting the estuary in a boat, she was sickened to find dead and dying fish everywhere, dead crayfish and crabs or those staggering about as their nervous systems were destroyed.

She was publicity shy and was aware of the potential downsides of challenging the million-dollar pesticide industry, but she was forced to conclude the worst effect of the chemical pesticides were their wider longer-term effects, including the inhibition of reproductive processes. Many species of birds were threatened with extinction, poisoned from eating fish that had eaten shrimp living in DDT-contaminated mud at the bottom of lakes. The peregrine falcons and iconic bald eagles were ceasing to breed on the US east coast, while in the Great Lakes region the eagles faced extinction because their egg shells were growing too thin, because DDT inhibits calcium production.

As insecticide residues were spreading on land, in rivers and into the oceans, they built up in food chains and, being persistent, were causing long-term toxic effects to humans.

From 1958 for almost four years Carson began meticulously researching her new book. She assembled information in ecological terms rather than simply portraying the effectiveness of the chemical pesticides. She did extensive research, citing dozens of scientific reports, conducting interviews with leading experts and reviewing materials across all disciplines.

Carson formed the view that governments and the industry were eager to deliver DDT's sweeping benefits, but they had promoted new pesticide technologies without knowing their full implications on the environment. When her book *Silent Spring* was published in 1962, it was a landmark in both changing the way chemical technology is introduced and initiating the modern environmental movement. Two years after her book was published, Rachel Carson died aged 56 of cancer. She had the satisfaction of knowing that her work was starting to influence policy.

Silent Spring was met with enormous public interest but substantial criticism from some scientists and chemical companies. After her warning of the threat to the environment from the indiscriminate use of more potent pesticides, her professional competence was challenged, but she had initiated the debate about their effect on the natural world and in food chains. She believed that life is much more interconnected and interdependent than was previously understood. She did not call for an outright ban on pesticides but urged caution, further studies and the development of biological alternatives. She wrote: *'...we have allowed these chemicals to be used with little or no advance investigation of their effect on soil, water, wildlife, and man himself. Future generations are unlikely to condone our lack of prudent concern for the integrity of the natural world that supports all life.'*

The book was to focus government attention on the threat posed by DDT, and it led to new policies to protect the environment and ultimately health and safety. Her ideas were influential to the 1963 US government enquiry on pesticides. It affirmed her call for limits on pesticide use and further research into their health hazards. Carson had articulated the dangers of improper pesticide use and the need for better pesticide controls. The public's adverse reaction to the once successful chemical technology of DDT brought with it a new way of regulating harmful pesticides.

In 1972 the US Environmental Protection Agency (EPA) prohibited most uses of DDT, and many other countries removed DDT from their lists of approved agricultural chemicals. It was seen to cause adverse environmental effects to wildlife and pose human health risks. It could accumulate in fatty human tissues and could also travel long distances in the upper atmosphere. In 2004 the UN Stockholm Convention outlawed many persistent organic pollutants (POPs) and restricted the use of DDT to solely controlling mosquito-borne malaria.

Controlling risks

Chemical technology has continued to produce important new pesticides, which have been successfully and safely deployed in light of the greater understanding brought by the experience of DDT. It was not, however, the only pesticide to cause controversy as a new generation of insecticides were developed in the 1980s known as neonicotinoids. One of these, Imidacloprid, became one of the most used insecticides in the world and works by disrupting the transmission of nerve impulses in insects, resulting in paralysis and death. Some ecological impact studies published within the past two decades have linked honey-bee colony collapse disorders with the use of neonicotinoids. Paraquat is one of the most effective, widely used non-selective contact herbicides, first manufactured by ICI in early 1962 and sold as *Gramoxone*. Paraquat was banned in Europe due to health concerns after 2007.

Today, the most widely used herbicide globally is glyphosate, sold as *Roundup*. It was developed in 1970 by a Monsanto chemist, John Franz, and is used extensively in agriculture, landscape management and gardening. It inhibits a plant enzyme integral to amino acid production, affecting the growing regions of the plants, killing plants in their growth cycle but not in their seed stage. It is sprayed on weed leaves and does not affect the soil. It has been used successfully for many years but recently there have been some studies into its possible health risks.

While the chemical industry continually develops new synthetic pesticides to make enhancements over the older products, regulations now put a higher priority on reducing the risks of pesticides in food and emphasize environmental protection. Some synthetic pesticides are still necessary as many uncontrolled pests cause serious problems, such as mosquitoes carrying serious diseases like West Nile virus or malaria; wasps or ants invading housing; domestic animals suffering harm and illness after infections by parasites or fleas. Invasive weeds spreading in wilderness areas can eliminate native plants. Pesticides are very often needed in food storage facilities to prevent rodents and insects damaging vital stored food, seed supplies and grain stocks.

Before a pesticide can be sold it must be approved by a government regulatory authority. This requires costly scientific studies to be conducted to determine whether the pesticide is effective, if it could be harmful to beneficial species or might cause adverse health effects in people over the longer term. Many of the concerns revealed from early pesticides have led to new farming practices that reduce pesticide runoff from crops to avoid polluting lakes, rivers and water supplies. Food companies have also been required to address consumer concerns by minimizing residues left on food. There is a particular focus on preventing pest resistance from developing by alternating using pesticides with different methods. Recently, in many parts of the world, pesticides are used in combination with alternative means of pest control such as crop breeding.

Farmers have been breeding animals and new varieties of plants for hundreds of years to enhance or avoid certain qualities. A growing new technology for controlling pests has been the use of genetically modified crops or organisms (GMOs). Genetic modification allows genes with specific attributes to be inserted into a plant or animal. In 1994 a genetically engineered soybean was approved that is resistant to glyphosate herbicide. This crop allowed glyphosate to control other competitive weed plants without endangering the valuable main crop. In 2005 a genetically modified version of golden rice was created to address vitamin A deficiency in the many populations that depend on rice as a staple food. The benefits of GMO crops include reductions in the use of chemical pesticides, enhancing nutritional value, creating more drought-tolerant varieties to cope with climate change and extending the storage life of agricultural products. The risks of GMOs have also to be identified, to ensure they have no long-term impact on human health or the environment.

Food security

More recently, crop protection efforts have evolved with new sustainable or 'regenerative' practices focused on using pest-resistant, drought-tolerant or herbicide-resistant crop varieties, soil improvement methods and companion planting. In the latter, different crops are planted in proximity for weed suppression, pest control, pollination or to provide habitat for beneficial insects.

Nonetheless, global consumption of pesticides stands at over 3.5 million tonnes annually and is valued at over $80 billion. Demand for pesticides will continue to rise given the need to feed a rapidly growing global population. This is combined with the threats to food security that are posed by climate change, greater urbanization leading to less land available for agricultural use and the continued emergence of new pests and plant diseases.

Q: Quantitative chemistry

Antoine-Laurent Lavoisier pioneering chemist

'When you can express something in numbers, you can know something about it' – Lord Kelvin

The success of many of the new chemical technologies in the past 200 years has relied in part on quantitative measurement of chemical reactions and processes. There were many pioneering chemists whose revolutionary ideas laid the foundations of not just modern chemistry but also many of the great innovations in chemical technology.

The emergence of modern chemistry dates from 1661 when Robert Boyle of the Royal Society in London published *The Sceptical Chymist* stating the concepts of the elements, of alkalis and acids and refuting the mystical ideas of alchemy. Boyle defined a chemical element as being *'certain primitive and simple, or perfectly unmingled bodies; which not being made of any other bodies, or of one another, are the ingredients of all those call'd perfectly mixt bodies are immediately compounded, and into which they are ultimately resolved'*. Soon discoveries of new elements and their compounds were made, although chemistry, unlike physics, which had advanced through the work of Isaac Newton, lacked a sound theoretical and quantitative basis until the early 19th century.

Henry Cavendish (1731–1810) from Britain was one of the greatest experimental and theoretical chemists of his age. He was distinguished for great precision in his research into the composition of atmospheric air, the discovery of hydrogen (inflammable air), the properties of gases, the synthesis of water, the law governing electrical attraction and repulsion, a mechanical theory of heat and even determination of the mass of the earth. Cavendish was the son of a British nobleman and politician. Notoriously shy, he entered Cambridge in 1749 but left without taking a degree, returning to work in his father's home, where he had his own laboratory built. Chemists were just beginning to recognize that the gases evolved in many chemical reactions were distinct entities, although their chemical character was not yet known. Cavendish conducted many experiments using the terminology of the 'phlogiston theory'. This theory was developed by the German scientist Georg Ernst Stahl in the early 18th century and offered a simple explanation of combustion: that every combustible substance contained a universal component of fire or phlogiston (Greek: *phlogistos* 'to set on fire'). Because a substance such as charcoal lost weight when it burned, Stahl reasoned that this change was due to the loss of its phlogiston to the air. The difficulty was when metals were strongly heated in air, the resulting oxides weighed more than the original metal, not less, as would be expected if the metal had lost its phlogiston component.

Cavendish produced *Three Papers Containing Experiments on Factitious Air* in 1766. These papers added greatly to knowledge of the reaction of alkalis to produce fixed air (carbon dioxide) and of dilute acids on metals to produce inflammable air (hydrogen). Another English scientist, Joseph Priestley, had noted that hydrogen when ignited left moisture on the sides of a previously dry vessel. In 1784 Cavendish determined the composition of this water, showing that it was a compound of oxygen and hydrogen. In 1785 Cavendish then carried out an investigation of the composition of atmospheric air, obtaining impressively accurate results. He used a device that measured the change in volume of a gas mixture of his own invention to obtain more accurate results rather than the inexact method of measuring gases by weighing them. He observed, besides nitrogen (78%) and oxygen (21%), there remained a volume of unknown gas amounting to 1/120th of the volume of the nitrogen. This was the small fraction of inert gases in the atmosphere. Cavendish's great experimental care and precision also allowed him to measure the density of hydrogen, which was impressive given the techniques then available.

Cavendish went on to show that fixed air (carbon dioxide), a product of reaction of chalk with acids was also the gas evolved from the fermentation of sugar. He was one of the first to investigate the conductivity of aqueous solutions and metals, then he examined the nature of heat. Thinking heat was the result of the motion of matter, his theory was both mathematical and mechanical; it contained the principle of the conservation of heat (or energy) and the concept of the mechanical equivalent of heat, which was later formalized by James Joule in 1845. If Cavendish had published all his ground-breaking work his influence would undoubtedly have been greater, but much of it only came to light a century later in his notes.

Pioneer

Another great pioneer of quantitative chemistry was also a meticulous experimenter, Antoine-Laurent Lavoisier (1743–1794). He established the law of conservation of mass, determined that combustion and respiration are caused by chemical reactions with oxygen and helped systematize chemical nomenclature.

The son of a wealthy Parisian lawyer, he entered Mazarin College in Paris when he was 11. His passion became science, although his father demanded he complete a law degree. After graduation, he began a long collaboration with geologist Jean-Étienne Guettard, who worked on a geological survey of France. Lavoisier showed an early aptitude for quantitative measurement and soon began applying his interest in chemistry while analysing geological samples. At age 25, he was elected to the elite French Academy of Sciences. By 1772 Lavoisier had turned his curiosity to the study of combustion. He became very sceptical of the theory of phlogiston. In his most famous experiment, he burned hydrogen gas in air to produce water. Lavoisier believed that water must be a combination of a substance in the air (which he later called oxygen) and hydrogen.

His intelligent wife, Marie Paulze, took a great interest in his scientific work, helping in the laboratory, translating and drawing sketches of his experiments. In 1775 Lavoisier was appointed to the Royal Gunpowder Commission to improve the production of gunpowder. He took up residence in the Paris Arsenal, where he equipped a fine laboratory and attracted young chemists from all over Europe to become part of the growing 'chemical revolution'. He succeeded in producing better gunpowder by increasing the supply and the purity of its constituents – saltpetre (potassium nitrate), sulfur and charcoal – as well as by improving the methods of granulating the powder. The young chemists in his laboratory said: '...*you heard this man with his precise mind, his clear intelligence, his high genius, the loftiness of his philosophical principles illuminating his conversation.*'

In his work he was systematic in determining the weights of reagents and products involved in chemical reactions, including the gaseous components, and believed matter would be conserved in any reaction (the law of conservation of mass). His work led him to conclude that combustion and respiration were being caused by chemical reactions with the component of the air he named 'oxygen' (from Greek for 'acid maker'). He established that water is made up of oxygen and hydrogen.

It was another British experimental chemist, Joseph Priestley (1733–1804), also working on the chemistry of gases, who had first discovered oxygen. Priestley was born into a family of successful wool-cloth makers in Yorkshire but was not permitted to attend university due to his religious beliefs. Between 1772 and 1790 he carried out intensive experiments on gases, designing ingenious apparatus, including a pneumatic trough to collect gases over mercury instead of water, allowing him to isolate and examine gases that were soluble in water and leading to his discovery of many new gases.

His reputation was founded upon the discovery he made in 1774, when he obtained a colourless gas by heating red mercuric oxide. Finding that a candle would burn in it and that a mouse could live in this gas, he called it *dephlogisticated air*. Priestley had found that *'air is not an elementary substance, but a composition* [mixture] *of gases'*. Like Cavendish, Priestley also abandoned the phlogistic theory. His religious views and his support for the American and French Revolutions forced him to flee Britain in 1794 for Pennsylvania in the US, where he continued his research work until his death.

In a meeting that was highly significant for the future of chemistry, Priestley had described to Lavoisier his discovery of the new gas. Lavoisier immediately repeated Priestley's experiments, giving him the understanding needed to develop his revolutionary theory of chemical reactions that would finally dispel the phlogiston theory and transform chemistry. Burning substances, Lavoisier argued, did not give off phlogiston at all; they reacted with Priestley's new gas, which Lavoisier had named oxygen.

Lavoisier considered 33 substances to be elements – by his definition, substances that chemical analyses had failed to break down into simpler entities. He published a textbook in 1789, *Traité élémentaire de chimie (Elements of Chemistry)*, which convinced European chemists of the validity of his theories and really began the quantitative study of chemistry. Unfortunately, this was at the start of the French Revolution and the country became increasingly chaotic, which prevented the progress of science. Lavoisier had previously bought a share in the private corporation that collected taxes for the French Crown and so was subjected to abuse from Jean-Paul Marat, the infamous revolutionary, who publicly denounced him for supporting his laboratory with income from a tax-collecting firm. All the members of the firm were arrested, including Lavoisier, despite his past services to France. He was convicted for being a tax collector for the king and executed by guillotine at the height of his success on 8th May 1794.

The famous mathematician Joseph-Louis Lagrange remarked on Lavoisier's tragic death: *'It took them only an instant to cut off that head, and a hundred years may not produce another like it.'*

Modern chemistry

Perhaps the most important and pioneering British chemist was John Dalton (1766–1844), whose ground-breaking 'law of partial pressures' would also enable quantification of many of the newly discovered chemical technologies. In 1808 he further transformed the understanding of chemistry with his chemical atomic theory, laying the foundations for all future quantitative work.

Dalton was born into a modest family, his father a weaver in Cumbria, England. While he received little formal education, he had a sharp and curious mind. Being a Quaker, he could not attend a traditional university but taught mathematics and natural philosophy at a dissenting college in Manchester. He joined the respected Manchester Literary and Philosophical Society, founded in 1781, which provided him with a stimulating intellectual environment and laboratory facilities. He wrote the first recorded description of colour blindness and kept daily weather records. He published papers on meteorological topics and gases. One contained Dalton's independent statement of Charles's law: that *'all elastic fluids expand the same quantity by heat'*. What began as his interest in meteorology would lead to a powerful and wide-ranging new approach to chemistry and eventually atomic theory. The emergence of new experimental techniques enabled Dalton to study the absorption of water vapour by air at different temperatures. In 1801 Dalton presented his research offering new insights into the nature of gases. He had discovered that the air is not a vast chemical solvent, as Antoine-Laurent Lavoisier had thought, but a mechanical system, where the pressure exerted by each gas in a mixture is independent of the pressure exerted by the other gases. His subsequent work on the constitution, evaporation and thermal expansion of gases led to his derivation of his 'law of partial pressures', which states that *'in a mixture of non-reacting gases, the total gas pressure is equal to the sum of the partial pressures of the individual gases'*. He claimed that the forces causing pressure acted only between atoms of the same kind and that the atoms in a mixture were indeed different in weight and complexity.

Dalton thought that all gases could be liquefied provided their temperature was sufficiently low and pressure sufficiently high. This was to inform his theories on the atom, his most influential work in chemistry. Dalton thought that atoms of different elements vary in size and mass. He proposed that each element had its own kind of atom, determining the relative masses of each different kind of atom, a process that could be accomplished by considering the number of atoms of each element present in different chemical compounds. The idea that there were multiple elements, each made up of its own, unique atoms, was new and controversial at the time. Since the time of the ancient Greeks, it had been thought that atoms of all kinds of matter are alike.

In a paper in 1803, he described his method of measuring the masses of various elements, including hydrogen, oxygen, carbon and nitrogen, according to the way they combined with fixed masses of each other in fixed proportions. Dalton's measurements allowed him to formulate the 'law of multiple proportions', commonly called Dalton's Law, which states that '*when two elements form more than one compound, the masses of one element that combine with a fixed mass of the other are in a ratio of small whole numbers*'. For example, Dalton found that 12 g of carbon could react with 16 g of oxygen to form carbon monoxide, but he also found that 12 g of carbon could react with 32 g of oxygen to form carbon dioxide. Dalton was intrigued that the ratio of the two (32:16) simplified to a ratio of 2:1. As elements combine in fixed, whole-number ratios to form compounds it seemed that compounds were made up of molecules containing two or more atoms of different elements. The problem remained, however, that a knowledge of the ratios was insufficient to determine the actual number of elemental atoms in each compound.

In 1808, when Dalton proposed his atomic theory (*New System of Chemical Philosophy*), it was the first successful attempt to understand the nature of chemical reactions and visualize how the elements combined. He used the term 'atom' from Greek *atomos* meaning 'cannot be divided'. According to Dalton's atomic theory, matter was made up of atoms, which were indivisible, unchangeable and indestructible building units. While an element's atoms were all the same size and mass, different elements possessed atoms of varying sizes and masses and underwent chemical reactions by joining with each other in fixed whole-number ratios to form compounds.

There was hard experimental evidence that quantitively explained reactions, but clearly atoms could not be seen. His atomic theory made it easier to express stoichiometric proportions in terms of atoms instead of absolute mass. Many chemists began to employ his atomic theory, as his concept of the atom became useful for explaining molecular structure in organic chemistry or the spacing and movement of molecules in gases. Dalton published a diagram numbering the elements: (1) for hydrogen, (2) nitrogen, (3) carbon, (4) oxygen, (5) phosphorus and so on. He suggested how molecules might be represented when the atoms combine to form compounds, say a molecule of water as 'OH' and ammonia as 'NH'. Although these were incorrect, his ideas laid the foundations for the development of modern quantitative chemistry.

John Dalton's elements and their combinations (1808)

His atomic theory gained widespread recognition and he is often considered as the 'father of modern chemistry'. When he died in 1844, such was his reputation that tens of thousands of local people paid their respects in Manchester

Town Hall before his burial. Two important merits of Dalton's atomic theory were that it did not violate the law of conservation of mass and it provided a basis to differentiate between elements and compounds. The theory would not explain how an element could have isotopes, such as deuterium in hydrogen, or allotropes, such as carbon with widely different forms such as diamond and graphite.

Chemistry's theoretical building blocks

After the start of modern quantitative chemistry by Cavendish, Presley, Lavoisier and Dalton there followed many more discoveries. The Italian chemist Amedeo Avogadro (1776–1856) built on Dalton's work by proposing that equal volumes of gases contain equal numbers of molecules (Avogadro's hypothesis) in 1811. There were also Carnot's work on establishment of the laws of thermodynamics (1824), Faraday's laws of electrochemistry (1833), Frankland's work on valency (1852) and the development of chemical thermodynamics by Gibbs and Van 't Hoff (1870).

New atomic theories evolved with more advanced experimental research. In 1904 British physicist Joseph J. Thomson discovered electrons and then proposed a theory of the divisible atom. His model is popularly referred to as the 'plum pudding model' with the electrons distributed uniformly in a sphere, like raisins in a plum pudding. He thought the charge of the positive sphere was equal to the negative charges of the electrons. New Zealand physicist Ernest Rutherford demonstrated the nuclear nature of atoms by measuring the deflection of alpha particles passing through a thin gold foil. In 1911 Rutherford formulated his 'nuclear' atomic model, in which a very small charged nucleus, containing most of the atom's mass, was orbited by electrons. He determined that the number of protons in an atom equals that of the electrons and that there are also neutral particles in the nucleus (later called neutrons). In 1913 Danish physicist Niels Bohr proposed a 'planetary' model, in which electrons revolved about the nucleus, like the planets orbit the sun. The electrons in 'orbit' had what Bohr termed 'constant energy'. When these particles absorbed energy and transitioned into a higher orbit, they became 'excited' electrons. When the electrons returned to their original orbit, they gave off this energy as electromagnetic radiation.

By the 1920s there were many scientists making contributions to atomic theory, including theoretical physicist Albert Einstein developing his theory of special relativity. The pioneers of quantum mechanics redefined how chemists fundamentally think about matter, particularly explaining the behaviour of electrons in terms of a 'wave' rather than a purely particle model. Instead of electrons orbiting a nucleus it was shown (mathematically) that the electrons exist as waves in discrete, uniquely shaped clouds around the nucleus called 'orbitals' where the electrons are most likely to be at any given time. Atomic orbitals are the basic building blocks of the electron cloud or wave mechanics model, the modern framework for visualizing the sub-atomic behaviour. Molecular orbital theory was developed in 1932 by US chemist Robert Mulliken (1896–1986) while Linus Pauling (1901–1994) advanced the new fields of quantum chemistry and molecular biology.

By the 1960s, as scientists investigated atoms with increasingly more powerful instruments, they discovered that the protons and neutrons that make up the nucleus are in turn made of even smaller particles called 'quarks', and since then many more sub-atomic and elementary particles have been discovered. Contemporary chemical technologists have increasingly been able to employ a wide range of quantitative tools and models when developing vital new technologies.

R: Recombinant DNA biotechnology

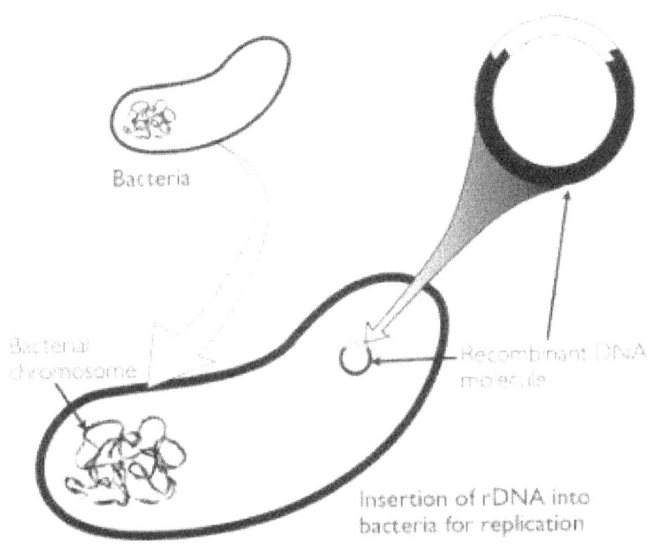

Insertion of recombinant DNA molecule into bacteria

New tools in the tool box

Chemical technologies have been very successful over the past few centuries but increasingly new biotechnologies such as recombinant DNA (rDNA) techniques promise to transform the future in many ways. 'Biotechnology' broadly refers to the use of cellular and living organisms or their components, including genes, to develop and manufacture products that improve human health, food or the environment, create energy resources or support fields like forensic science. On an industrial scale many of the skills and techniques that are used in biotechnology have been built on those already used in chemical manufacturing.

A new biotechnology industry has grown up harnessing cellular and biomolecular processes allowing advances and adaptations in numerous fields. Traditional biotechnology, with processes such as fermentation, uses living organisms in their natural form, but now it is possible to modify their genetic make-up using powerful techniques such as recombinant DNA technology. Here, genetic material from one organism is artificially introduced into the genome of another organism and then replicated and expressed by that other organism. This was made possible through the work of Herbert Boyer, Stanley Cohen and Paul Berg in the 1970s. Recombinant DNA technology has proven important to the production of vaccines and protein therapies such as human insulin, interferon and human growth hormone and in the development of gene therapies.

Industrial biotechnology has a long history in the UK, from the production of ethanol by fermentation during World War I, when the bacterium *Clostridium acetobutylicum* was fed with starch from potatoes and corn to produce a mixture of propanone (acetone), butanol and ethanol. The propanone was required to produce munitions, although this route became uneconomical with the rise of the petrochemical industry, which provided much cheaper feedstocks. Then, after Scottish researcher Alexander Fleming identified the antibiotic penicillin in 1928, the US company Pfizer which was already successfully making citric acid using fermentation, developed large-scale life-saving penicillin production (see Y).

The modern era of biotechnology is linked to several scientific breakthroughs in understanding genetics. In 1909 the term 'gene' was first employed by Wilhelm Johannsen (1857–1927), who described the gene as a carrier of heredity and hereditary disorders. DNA was shown to be the repository of genetic information. The discovery of its molecular structure by James Watson and Francis Crick in 1953 was another critical step forward in the understanding of genetics, but it was another two decades before it was possible to take a gene from one organism and insert it into the DNA of another. This was the development of recombinant DNA biotechnology (rDNA).

Pioneers

In 1971 Paul Berg at Stanford University, California, had demonstrated the feasibility of splicing and recombining genes for the first time. The next landmark was the insertion of recombinant DNA into bacteria in such a way that the foreign DNA would replicate naturally. This was accomplished by Herbert Boyer at the University of California, San Francisco, in collaboration with Stanley Cohen of Stanford. Boyer was born in 1936 in Pennsylvania, where he attended college to undertake premedical studies. He was soon captivated by research and chose to major in chemistry and biology. He completed a PhD in biochemistry at the University of Pittsburgh and a postdoctoral fellowship at Yale. In 1966 he accepted an appointment at the University of California, San Francisco, which was becoming a centre for the emerging field of biotechnology.

He realized that the enzyme *EcoRI* cut DNA in such a way that the ends were not blunt but staggered, so that no molecular additions were needed to make one severed piece latch on to another piece possessing complementary cuts.

Boyer began an effort to create rDNA to insert in the bacterium *Escherichia coli* (*E. coli*) by trying to use the enzyme to 'open up' the DNA of a bacterial virus. He became frustrated, however, when the enzyme cut the DNA in five places instead of one. In November 1972 Boyer met Stanley Cohen at a meeting on plasmids. A plasmid is a small DNA molecule, found especially in bacteria, that is physically separate from and can replicate independently of the bacterium's chromosomal DNA. Cohen was born in 1935 in New Jersey, was a graduate of Rutgers University and took medicine at the University of Pennsylvania. He began his career in medical research working in New York on the complex mechanisms that control gene expression in the bacterial virus lambda. In 1968 he had joined the Stanford University School of Medicine.

While Boyer was describing his data showing the nature of the DNA ends generated by EcoRI cleavage, Cohen reported on a procedure recently discovered in his laboratory that enabled bacteria to take up plasmid DNA and produce offspring that contained self-replicating plasmids identical to the original implant-clones. Late one night at the conference, the two researchers decided to collaborate on a project to discover what genes are present on plasmids and how they are arranged. The collaboration meant plasmids isolated at Stanford were transported to Boyer's laboratory in San Francisco for cutting by EcoRI and then analysis of the DNA fragments. These were transported back to Stanford, where they were joined and introduced into *E. coli*, where they multiplied. Then the brand-new recombinant plasmids were isolated and analysed in each laboratory.

They had success in the spring of 1973 when one of Cohen's plasmids (pSC101) could make *E. coli* bacteria resistant to the antibiotic tetracycline. Boyer and Cohen soon moved on to more complicated cloning experiments. They joined tetracycline-resistant plasmids with kanamycin antibiotic-resistant plasmids and inserted them in *E. coli*. Next, disproving a long-held belief, they showed that genetic materials could indeed be transferred between species. They cut a piece of staphylococcus plasmid (the bacterium responsible for many infections), spliced it with one of the many *E. coli* plasmids and inserted the whole in *E. coli*. The DNA from the staphylococcus, a different species of bacterium, was successfully propagated in *E. coli*. An even greater demonstration of interspecies cloning was the insertion into *E. coli* of genes taken from the South African clawed frog.

In 1974 Cohen and Boyer developed a technique for splicing together strands of DNA from more than one organism. Their work would facilitate the economic manufacture of human proteins at a larger scale, which allowed therapeutic uses. Commercial ventures were to follow soon, exploiting the new rDNA technology. At the forefront of these was the pioneering biotechnology company Genentech, which Boyer founded in 1976 with a young venture capitalist, Robert Swanson.

Boyer and Cohen, and other scientists involved in cloning experimentation, soon recognized the feasibility of using bacteria into which human genetic information was incorporated to duplicate the body's natural means of fighting disease and to remedy birth disorders. In 1977, even before Genentech had its own facilities, Boyer working with Keiichi Itakura, an organic chemist at a medical centre in Duarte, California, succeeded in expressing a mammalian protein in bacteria. This involved somatostatin, a hormone that stops the release of other hormones that regulate bodily functions and processes. Recombinant somatostatin was shown to be virtually identical to the naturally occurring substance. It was then in 1978 that Boyer and Itakura went on to construct a plasmid that coded for the human hormone insulin. Insulin had been used in treating type 1 diabetes for a half century but it was extracted from animals and so had a slightly different form from human insulin. Producing synthetic human insulin would satisfy the high demand for a vital medicine, and its regulatory approval would be relatively straightforward. The recombinant insulin, known as *Humulin*, was produced by Genentech working in a joint venture with Eli Lilly.

Humulin became the first modern biotechnology product to appear on the market, to be followed by many new drugs including the first recombinant DNA vaccine for livestock. Genentech had not only demonstrated that a major animal-derived drug like insulin could be mass-produced through the cloning and expressing of significant genes in bacteria, but they were also pioneering the commercialization of recombinant DNA technology, which drove a whole new era for the biotechnology industry based on genetic engineering.

In the next 20 years, the new 'biotech' industry grew to produce almost 100 recombinant DNA products for treating disease or for vaccination. Boyer built up Genentech from an initial investment of $500 to a value of about $40 million in just over 10 years. It was a remarkable success for innovative biotechnology, supported by the US entrepreneurial culture. Such pioneering businesses working with major pharmaceutical companies, while retaining valuable intellectual property rights, became the dominant model for the next generation of biotech entrepreneurs.

Another company, Biogen, produced an interferon cancer treatment using recombinant DNA, while the possibility of treating HIV/AIDS generated further enthusiasm for the new technology. In addition to synthetic insulin, several proteins were genetically engineered and approved as medicines including human growth hormone, alpha-interferon, a hepatitis B vaccine and a protein for treating blood clots. In agriculture, the rDNA technology was used to produce the first genetically engineered crop (GMO), a pest-resistant maize known as *Bt corn* with improved resistance to pests, both increasing crop yields and reducing the use of chemical pesticides so leading to environmental benefits.

New techniques

In 1983 Kary Mullis, a biochemist at Cetus Corporation, California, developed the *polymerase chain reaction* (PCR), which allows a piece of DNA to be repeatedly replicated. The process works by heating double-stranded DNA so that it separates into two single strands. Then primers are attached to the individual DNA strands. Enzymes called DNA polymerases attach at the primer sites and replicate the rest of the strand of DNA. This process can be repeated many times, with each iteration doubling the number of exact DNA copies. Mullis was awarded the Nobel Prize for Chemistry for his work in 1993. The ability to replicate DNA opened doors in many fields. It allows forensic scientists to apply genetic techniques even if there is only a small amount of genetic material found at a crime scene. In medicine, it helps to identify the cause of infections. In research, it is an essential technique used during the sequencing of genomes. PCR is now a ubiquitous technique in biochemistry laboratories for research and new product development worldwide. A PCR-based test was important in the rapid testing for the COVID-19 virus during the 2020 pandemic.

In 2001 the sequence of the entire human genome was published, making it possible for researchers to begin identifying new treatments. In 2004 the UN Food and Agriculture Organization endorsed the use of crops derived from biotechnology, considering them a complementary tool to traditional farming methods with the potential to help poorer farmers and consumers in developing nations.

In 2012 a team of scientists led by Jennifer Doudna at University California, Berkeley, and Emmanuelle Charpentier at University of Umea in Sweden announced a new fast and precise method for editing small pieces of the genetic code. The so-called CRISPR system takes advantage of a defence strategy used by bacteria. Its rather unwieldy name stands for 'clustered regularly interspaced short palindromic repeats' – a family of DNA sequences found in the genomes of single-celled organisms such as bacteria. The technique allows for the highly specific and rapid modification of DNA in a genome, the complete set of genetic instructions in an organism.

It is possible to carry out genetic engineering on an unprecedented scale at a very low cost allowing for the introduction or removal of more than one gene at a time. The new technique is already being explored for a wide number of applications in fields ranging from agriculture to designing new cereals, vegetables and fruits. In human health, it could pave the way for the development of new treatments for rare metabolic or neurological disorders and genetic diseases such as haemophilia, cystic fibrosis or Huntingdon's disease. It is also being utilized in the creation of transgenic animals to produce organs for transplants into human patients. It may be possible to genetically engineer insects to eliminate insect-borne diseases such as malaria, transmitted by mosquitoes, or Lyme disease, transmitted by ticks. The importance of the CRISPR technique was recognized by the awarding of the 2020 Nobel Prize in Chemistry jointly to Doudna and Charpentier.

Jennifer Doudna holds model of CRISPR-Cas9

Balancing the risks and rewards

As with some of the older chemical technologies, the discovery of recombinant DNA biotechnology and advances in genetic engineering could have major human and societal consequences. The many advances in genetic engineering and biotechnology have initiated a major ethical debate about how the new technologies should be used. Safety is the primary concern to ensure that changes made in one part of the genome do not lead to unforeseen harmful consequences for humans or the environment; but these are balanced with a potentially fantastic ability to treat serious diseases, increase food supplies for a growing population and provide benefits to the wider environment.

Potentially, there are now some very effective new tools to address many of the problems faced globally.

S: Soap

Soap packages by Lever Bros.

Cleaning up

Soap is the product of one of the earliest chemical technologies and perhaps the first manufactured product people encounter. The soaps and detergents made in the chemical industry are used to produce cleaning products for personal hygiene, sanitation, laundry and household uses. It has been said that no industry has had a lowlier origin than the manufacture of soap, nor is there one that so intimately has reflected progress in the refinements of life.

Soap is today considered a daily necessity. Few consumer products are more ubiquitous than soap, made as many well-known and long-established brands such as *Sunlight*, *Pears*, *Palmolive*, *Zest*, *Dove* and *Olay*. Soap has over centuries been created from countless variations of ingredients but essentially it has two main components – oils or fats and an alkali that enables the hydrolysis of esters. The benefits of soap came to wider appeal in Europe from the 17th century, as maintaining cleanliness was part of an improved quality of life. Advancements in chemical technology enabled soap to become more effective in cleaning, to find medicinal uses and to be used as surface active agents. Several great Victorian entrepreneurs drove the industrialization of soap production. Foremost of these was William Lever (1851–1925), who built Britain's largest soap and consumer products company. Lever based his original soap, marketed under the *Sunlight* brand, on palm kernel oil, cottonseed oil, resin and tallow. In 1907 the German company Henkel put the first modern washing powder called *Persil* on the market, which combined detergent, bleach and disinfectant in one product. By the late 19th century, the widespread adoption of soap and clean water had begun to transform public health.

Although used for cleaning, soap is of utmost importance in reducing the prevalence and spread of infectious diseases. Diseases spread by unwashed hands include typhoid, cholera, polio, hepatitis and norovirus, the latter being the most common cause of viral gastroenteritis in humans. Simply cleaning hands before preparing food or eating, undertaking medical procedures or after visiting the toilet can often save lives by reducing the spread of illnesses. Poor hygiene or sanitation in these situations can lead to transmission of disease pathogens as infection passes from the hands of one person to the mouth (or food) of another. Some infections spread quickly through large groups of people in close quarters, such as in hospitals, cruise ships, student accommodation and care homes. Illnesses such as the common cold and influenza, chicken pox, meningitis and streptococcal infections can spread through droplets in the air, which land on nearby objects and when touched transmit germs through the nose, mouth or eyes. A hospital-acquired infection (nosocomial infection) can be acquired in hospital or any healthcare facility and be potentially fatal and difficult to eradicate. Examples include *methicillin-resistant staphylococcus aureus* (MRSA) and *Clostridium difficile* (C. diff.). Many are transmitted to patients from other patients or medical staff following poor hand washing practices.

Recent pandemics and viral epidemics such as the COVID-19 pandemic in 2020 or serious seasonal flu outbreaks have created an ever-increasing demand for disinfecting products – of which simple soap-based products remain a vital component. The UN estimates that some 10% of people worldwide have inadequate handwashing facilities, which results in half a million people dying each year from infections that could have been prevented with good hand hygiene using soap.

History

The discovery of soap predates recorded history but excavations near ancient Babylon provide evidence for soapmaking around 3000 BC using a slurry of ashes and water to remove grease from raw wool and cloth prior to dying. It was realized that a little grease improved the performance of the alkali, giving rise to making soap solutions directly by boiling fats and oils in the alkali before using it for cleaning. A clay tablet from Mesopotamia dating to 2200 BC had a soap recipe inscribed on it that used animal fat and tree ash. Egyptian manuscripts from 1500 BC describe a substance made by combining animal fats and vegetable oils to create a soap-like base. Around 200 AD the Greeks had combined lye and ashes to clean their pots and statues. The Gauls and Romans also used animal fat, beech tree ashes and goat tallow to produce both hard and soft soap products.

During the Roman period, with the popularity of public and private baths across the empire, essential oils and abrasive materials were used with a strigil, a metal implement for scraping the skin free of oil and dirt. The Romans also made use of a type of clay found near Rome called 'sapo', from which the word soap may be derived, although they may have first observed the art of soapmaking, using animal fats and plant ashes, by the Celts, who called it 'saipo'. The remains of a soap factory were recently discovered in Pompeii, which was covered up by the eruption of Vesuvius in 79 AD. Roman medicinal books indicate that the use of soap was beneficial for health and long life. The boiled leaves or roots of some perennial plants such as 'soapwort' (*Saponaria officinalis*) were used as a gentle soap to clean delicate textiles. The use of soap in personal hygiene and for shampoo is mentioned in the second century by the Greek physician Claudius Galen. The trade of a 'soap-boiler' is recorded in commercial records. After the fall of the Romans, the use of soap declined during the early Middle Ages, when personal and domestic hygiene were probably a lower priority as life was turbulent and precarious and disease was common, although in Asia hygiene remained enforced by tradition. The manufacture of soap in the Mediterranean region re-emerged by the end of the first millennium. Soaps largely were being used in the cloth industry rather than for personal hygiene, to prepare wool for dyeing, in towns such as Marseilles in France and Savona in Italy (the possible origin of *savon* in French). Secret recipes were handed down from master to apprentice.

In Britain the first written mention of soap appears in 1192 when Richard of Devizes referred to soap makers in Bristol and the unpleasant smells that their activities produced. By the 13th century, soapmaking in Britain had become centred in larger towns like Bristol, Coventry and London, each making its own variety. Large areas of woodland were harvested to meet the growing demand for wood ash. The production of soap in London was taking place in the 15th century on 'Soper's Lanes' in Bishopsgate and Cheapside. Throughout the Middle Ages soapmaking involved boiling olive oil (in Mediterranean countries) or animal fats (in Northern Europe) with an extract of plant ashes and lime. For example, in Spain, the plant salsola (saltwort) was burned to produce an alkaline ash called barilla. The locally available olive oil offered a good-quality soap made by salting out (graining) the boiled liquor with brine, allowing the soap to float to the surface, leaving the alkali (or lye), vegetable matter and impurities to settle out. This produced the first white hard soap. Plentiful supplies of high-quality olive oil and barilla ashes made regions like Castile in Spain and Marseilles in France renowned for the quality of the 'Castile' soaps they produced. Castile eventually became a generic name for hard white olive soaps.

In Britain production of soap was usually based on rendered animal fat, such as tallow from beef or mutton. By the 16th century three broad varieties of soap were available: coarse soap made from train oil extracted from whale blubber, sweet soap from olive oil and speckled soap made from tallow. The use of tallow to make soaps was discouraged as its use drove up the cost of candles. As imports of oils such as olive, palm, coconut, linseed and cottonseed grew, seaports such as London and Bristol were favoured. Italian soap makers improved soap by the introduction of perfumes. The emergence of soap as a consumer product allowed it to become a source of taxation. In 1632 King Charles I granted a monopoly to the Society of Soapmakers of Westminster to produce soap in return for tax payments of £4 per tonne. The government decreed that there should be no soap manufacture outside a one-mile limit of London and Bristol and set annual production quotas. In 1712 Queen Anne's government introduced a tax of 3 pence per pound of soap. All soap manufacturing pans were required to be fitted with a padlock, only to be opened by the exciseman who attended each soap boiling to avoid fraudulent accounting.

The Industrial Revolution brought steam power, iron vessels and new machine technologies, leading to greater soap production efficiency, economies of scale and better controlled conditions. Domestically, the combination of better soaps and advances in plumbing, running water and drainable baths began to make bathing the social norm. As the soap industry started to understand the chemistry involved, further efficiency was achieved and a wider variety of more fragrant, milder and colourful soaps was made to appeal to consumers. The growing tide of Victorian concern about cleanliness led the British government to abolish soap duty in 1852. Soap businesses became even more profitable as a result of Nobel's invention of dynamite (see G), which allowed the explosive nitroglycerine to be derived from glycerine, previously a waste product of soapmaking.

Chemistry of soap

Throughout its long history the chemical process for soap production has not fundamentally changed, although the chemistry of soap was not understood; rather, it was achieved by experimentation. Oils or fats are heated with alkali in a reaction that produces soap and glycerine. Fats and oils are glycerides of organic acids. Early soap production relied on animal fats and ash, produced by burning wood, as a makeshift source of alkali. The quality of soap is very dependent on the quality of the materials employed in the reaction. Modern soaps are sodium or potassium salts of long-chain fatty acids made by reacting either sodium hydroxide or potassium hydroxide with the fatty acids, derived from a variety of natural or synthetic fats, in a chemical reaction called *saponification*.

The French chemist Michel Eugene Chevreul (1786–1889) studied the chemistry of soap production in 1811, showing that the alkali split the fat into an alcohol, which he named glycerine (propane-1,2,3-triol), and the soap, which was the salt of a fatty acid. This knowledge paved the way for the great expansion of soap manufacture later in the century, which required higher-quality sources of alkali.

Soap is capable of cleaning dirty, greasy or oily items with water – normally water does not mix with grease or oil and the dirt adheres. The oils and water do not mix as the water molecules are *polar*, and the *nonpolar* oil molecules remain apart ('like dissolves like'). Soap and synthetic detergents are 'surface active' substances or *surfactants*, which reduce the surface tension of water by allowing it to be more effective in wetting and facilitating the cleaning of contaminated materials and surfaces. The soap molecules have a two-part chemical structure: a long hydrophobic ('water-fearing') hydrocarbon 'tail' and a hydrophilic ('water-loving') anionic 'head' $[CH_3(CH_2)_n COO^-]$. Here the length of the hydrocarbon chain (n) varies with the type of fat or oil used. The negative or anionic charge on the carboxylate $[COO^-]$ head is usually balanced by either a positively charged potassium $[K^+]$ or sodium $[Na^+]$ cation. The polar part mixes with the water while the nonpolar part mixes with the oil or grease because the carboxylic head, which is hydrophilic, interacts with water molecules due to ion-dipole interactions and hydrogen bonding. The nonpolar hydrophobic tails of soap molecules are embedded into greasy materials adhering to surfaces. These hydrocarbon chains are attracted to each other by dispersion forces and cluster together forming structures called *micelles*. In a micelle, the tails of the soap molecules are oriented into the grease or oil, while the heads face outward into the water, resulting in an emulsion of soapy grease particles suspended in the water. With agitation, the micelles are dispersed into the water and freed from the previously dirty surface.

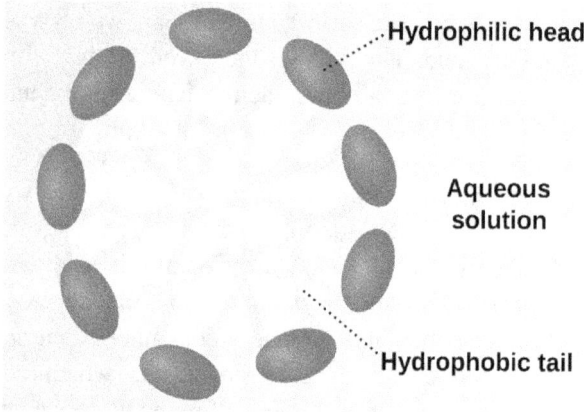

Cleaning grease with soap molecules

Soap production

When making soap, triglycerides contained in fat or oils are heated in the presence of a strong alkali, producing three molecules of soap for every molecule of glycerol. In the first part of this saponification reaction, sodium hydroxide is mixed with water to exothermically generate sodium ions and hydroxide ions. The triglyceride is broken down into fatty acids and glycerol through a two-step process called steam hydrolysis. This yields a fatty acid without its salt as well as glycerine (glycerol). Then, the sodium ions react with the fatty acids to form a fatty acid salt, which is the soap and water. The saponification reaction is endothermic and control of the temperature is important. If too high, the soap will separate during the hotter gel phase when the soap is in a mould. Perfume is carefully added to the soap, which hardens in the cooling and hardening phase. If the reaction takes place at a temperature higher than 120°C, the raw soap will saponify too quickly, becoming thick and making it difficult to pour into moulds.

The two common methods for producing soap are the cold and hot processes. The cold process for making soap takes 18 to 24 hours to complete the saponification. The soap from the cold process requires some weeks to cure, during which it loses water to eventually become hard. Cold process soap tends to be of higher quality, and the remaining glycerine from the saponification reaction is usually added to the soap as a natural skin softener. The hot process requires only two hours for the saponification reaction, because it is reheated in a double boiler, and hardening requires only one week.

Traditionally, industrial soap manufacturing involved the boiling of oils and fats with caustic solution in large open pans followed by the addition of salt or brine ('salting out' process), in which the soap separated from the alkali (or lye). A skilled operator would control the process by 'trowelling' to judge whether more brine or caustic was required and when the batch was ready for settling. By successive washing in brine, the alkali was separated from the soap and the glycerine recovered. The soap was dried and cut into bars for supply to the wholesale and retail trade. The shopkeeper cut the blocks into individual bars using a 'cheese' wire or sharp knife, then wrapped them in paper for the customers. As industrial capability evolved, soap was sold at pharmacists as it found applications in making pills, lotions and liniments, toothpaste, enemas and poultices.

An early industrial manufacturer of quality soap was Andrew Pears from Cornwall, who had a barber's shop in fashionable Soho. He recognized the need for a purer, gentler soap that would be kinder to the delicate alabaster complexions favoured in high society. In 1789 he commenced production of a new transparent soap at a factory in Wells Street, near Oxford Street. His *Pears* soap was produced by prolonged evaporation and drying from an alcoholic liquid soap in a process taking up to three months. The characteristic concave shape of the soap tablet was achieved not by moulding but by shrinkage in the drying process. The manufacturing process he perfected remains substantially unchanged to this day. The Pears pioneering advertising campaign used the oil painting *Bubbles* by Sir John Everett Millais, propelling Pears transparent soap sales worldwide. From the early 1800s the Cook family in Norfolk ran a large soapmaking company. In the US, two men from England, William Proctor and James Gamble, established their famous soap firm in 1839 at Cincinnati, Ohio while the B.J. Johnson Soap Company of Wisconsin made the popular soap *Palmolive* entirely from palm and olive oils. The most important pioneer of industrialized soap manufacturing was William Hesketh Lever.

Pioneer

William Lever (1851–1925) was born in a large family in Bolton, where his father ran a successful grocery business.

He left school at the age of 16 and joined the family business. In 1874 he married Elizabeth Hulme from his church, who influenced his values in business. William and his brother James decided to enter the soap business by buying a small, unprofitable soap works in Warrington in 1885. They formed a partnership with William Watson, a chemist from Bolton whose newly invented soap used imported palm oil, rather than tallow fat, to produce a free-lathering soap, which proved popular. Lever Brothers sold the soap at first as *Honey* and then as *Sunlight* soap. Watson's marketing acumen drove sales internationally. By 1887 the business was thriving and was producing 450 tonnes per week. Lever Brothers began looking for a new site on which to expand the soapmaking business. The company bought some flat marshy land in Cheshire, south of the River Mersey. The site was large enough to allow space for expansion and was a prime location between the river and a railway line. This became Port Sunlight, which grew to cover over 160 hectares with a new works and a model village to house the employees. It was similar in concept to the village that chocolate maker George Cadbury would build at Bourneville. Port Sunlight had facilities for education, recreation, gardening and entertainment of the workforce, including clubs for art, literature, science, crafts, music and sport. By 1914 there were 3,500 people living in 800 homes with a dedicated concert hall, theatre, library, gymnasium, art gallery, cottage hospital, open-air swimming pool, church, allotments and temperance hotel. The Levers soon instituted other employee benefits, including pensions, medical care, unemployment payments, profit sharing and free insurance. In return they required that employees observe a strict ethical code and the cottages were 'tied' to employment so had to be given up on leaving or breaching the codes.

By 1900 weekly soap output exceeded 5,000 tonnes and their successful *Sunlight* soap had been joined by products such as *Lifebuoy*, *Lux* and *Vim*, with many subsidiaries set up in the US, Switzerland, Canada, Australia and Germany. To expand further, the company cultivated its own palm oil plantations. The availability of alkali on a large scale was made possible by Leblanc's process for sodium carbonate (see A) and then the Castner–Kellner Alkali Company at Runcorn. During World War I, Lever Brothers undertook the manufacture of margarine and soon gained a dominant market position. It became Britain's largest manufacturing and exporting company. Lever's methods were noticeably different from normal British practice, building up a great organization that reflected his values and breadth of vision. Lever felt that his cheaper products drove the power of consumption by workers everywhere, which in turn supported increased production, leading to additional investment of capital, machinery and in turn more jobs for workers.

Lever built the international firm of Lever Brothers by consolidating many smaller companies into his multinational operation. These Soapmakers included William Gossage's firm, which made low-price soap brands such as *Rinso* and *Omo* at a factory situated by the Mersey, R.S. Hudson's soap powder factory at Liverpool and A&F Pears Ltd. Lever became a Member of Parliament and was raised to the peerage, becoming 1st Viscount Leverhulme of the Western Isles. By the time of his death in 1925 in London, aged 74, he was recognized as a leading figure in the late Victorian industrial and consumer revolution.

In 1929 the Unilever company was formed by a merger of Lever Brothers with the Dutch company Margarine Unie. It remains one of the UK's largest companies with 130,000 employees globally and a turnover over £50 billion. It is a multi-divisional international business with hundreds of consumer brands covering food, home and personal care products.

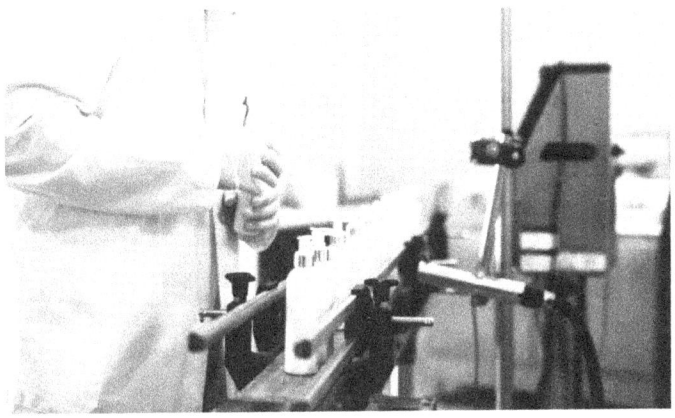

Liquid soap filling line

Healthy living

Soap manufacture today uses continuous processes with automated control systems in a science-led, highly competitive, multibillion-pound consumer-focused industry, whose products are a long way from the crude, evil-smelling soaps of the Middle Ages. Advances in chemical technology have enabled the creation of shower gels and specialized liquid cleaning products. Although bar soap is relatively low in cost, it is less effective when used in hard-water areas, whose water contains calcium and magnesium ions, which produce insoluble salts of soap that create an unpleasant soap scum. Many traditional soaps have been largely replaced in modern cleaners by synthetic detergents that have a sulfonate group $[R-S(=O)_2-O^-]$ instead of the carboxylate head, as these detergents tend not to precipitate in hard water and are generally better wetting agents. Liquid soaps and shower gels do not contain saponified oil but are mostly based on petroleum oil. They leave no residues on the skin or bathroom ware, are less irritant and contain conditioning agents and moisturizers to make them suited for both body and hair uses.

The size of the global soap and detergent market exceeds $170 billion and is continuing to grow as populations increase. About three-quarters of all the soap produced worldwide is used for personal hygiene. In wealthy countries there is increasing demand for new types of soap for specialist skincare, antibacterial or deodorant uses and for 'natural' formulations that claim to be eco-friendly. The availability of substitute products that can be used instead of bar soap, such as liquid soap or cleaning wipes, is forecast to grow faster in these countries.

However, in poorer countries it is estimated that half of child undernutrition is caused by lack of access to clean water, poor sanitation and inadequate hygiene leading to exposure to harmful bacteria. Handwashing with the age-old chemical technology of bar soap remains a cost-effective way to prevent health problems such as diarrhoeal diseases or emerging global diseases.

T: Tyres of synthetic rubber

Rubber car tyres made by Continental

Where the rubber hits the road

Tyres are thick rings of rubber, usually inflated, that are placed around vehicle wheels to provide traction in different conditions and allow safe and comfortable travel on roads. As is often the case, much of human economic and technological progress has depended on the development of new materials such as cement, iron and steel, plastics and synthetic rubber. Tyres made using the chemical technology of synthetic rubber have transformed the modern world of transport.

Chemically, natural rubber, known also as latex or caoutchouc, consists of polymers of unsaturated isoprene

(2-methyl-1,3-butadiene), mainly poly-*cis*-isoprene, with minor impurities of other organic compounds. In 1844 American Charles Goodyear first patented a way to cure natural rubber to increase its utility. Until the 1940s most tyres were made from natural rubber from Asia, but that source was no longer available during World War II. Tyre companies then developed ways to produce synthetic rubbers to meet the demand. The new synthetic rubbers were superior to natural rubbers in several respects, making use of plentiful oil-based raw materials, their thermal stability and resistance to oils or oxidizing agents. These synthetic rubber tyres revolutionized transport, facilitating the widespread use of cars, trucks, bicycles and aircraft.

History

Natural rubber was known to the indigenous peoples of the Americas, from which they made balls, bands and containers. It had been used within the Aztec empire and by the Maya Indians. Explorer Christopher Columbus (1451–1506) observed people waterproofing moccasins with tree gum. When first introduced into Europe rubber was used as an eraser ('India rubber'). The Portuguese used it to make containers for replacing the leather borrachas that were used to ship wine.

Natural rubber is obtained from the latex fluid secreted by certain plants, such as the rubber tree plant (*Hevea brasiliensis*). The latex is produced by specialized cells called laticifers, and plant scientists believe these to be an adaptation that protects certain species from insect attack. Natural latex has been obtained on commercial rubber plantations since the mid-19th century. It is harvested by hand each day by workers known as 'tappers' who insert a metal tap into each rubber tree, below which are hung collection cups. As the latex dries, the isoprene molecules crowd together, with one isoprene molecule attacking a carbon–carbon double bond of a neighbouring molecule. When a double bond breaks, the electrons rearrange to form a bond between the two isoprene molecules. The process continues until there are long strands of many isoprene molecules linked in a chain. These strands are the polyisoprene polymer. As the drying continues, the polyisoprene strands form electrostatic bonds holding the rubber fibres together, which allows them to stretch and to recover. Natural rubber is a thermoplastic with low tensile strength and low elasticity because of its molecular structure consisting of a mixture of polymeric chains with varying lengths.

The first major commercial use for rubber was on hot-air balloon fabric coated with rubber dissolved in turpentine. In 1820 a Scottish chemist, Charles Macintosh, cemented two pieces of fabric together with natural rubber that was made soluble by the action of the naphtha producing a waterproof garment commonly known as a *Mackintosh*. Rubber was widely used in shoes, boots, clothing, waterproofing, vibration damping, carpet underlays, cable insulation, bowling balls and protective gloves. Rubber articles still possessed serious defects; they became sticky in heat, and at low temperatures lost their elasticity. The properties of natural rubber can be altered by cross-linking the polymer chains through the process of vulcanization to make it more durable.

Pioneer

It was the American inventor Charles Goodyear (1800–1860) from Connecticut that discovered a way to cure natural rubber to increase its durability. In 1838 Goodyear met Nathaniel Hayward, who had treated rubber sheets with a solution of sulfur and turpentine and then dried them in the sun making them harder and more durable. Hayward patented the process he called *solarization*. Goodyear purchased the rights to use solarization and began experimenting with sulfur compounds, mixing latex rubber with sulphur. While carrying out one experiment, he accidentally dropped some rubber mixed with sulfur in a pan on a hot stove and found the rubber had transformed into a hard, durable but flexible material. He worked out a controlled process for this hardening, which he called *vulcanization* (from *Vulcan*: Roman god of fire and the forge). He found that varying the amount of sulfur changed the rubber's characteristics: the more sulfur, the harder it became. Goodyear's final process combined latex rubber, sulfur and lead oxide in high-pressure steam for up to six hours. He obtained a patent and by 1844 was producing the rubber on a more industrial scale.

The vulcanization process forms cross-links between sections of polymer chain, which results in increased rigidity and durability, as well as other changes in the mechanical and electrical properties of the material. When polyisoprene strands are heated with sulphur (typically 2-3%) and lead oxide, the sulfur atoms attack the double bonds in the polyisoprene strands and bind to the carbon atoms. Sulfur atoms can also form bonds among themselves (disulfide bonds) and cross-link adjacent polyisoprene strands to form a net-like structure in the rubber, which strengthens the rubber to make it harder, flexible and more durable. The number of sulfur atoms in the cross-link strongly influences the properties of the final rubber. Vulcanization is normally referred to as 'curing' and is generally irreversible. The attractive physical properties of vulcanized rubber revolutionized its application.

In the same year that Goodyear obtained his patent, Thomas Hancock (1786–1865), a British inventor, also filed a patent for the vulcanization of rubber, having studied Goodyear's process. This resulted in years of litigation, leaving Goodyear impoverished prior to his death in 1860. The Goodyear Tyre & Rubber company in Ohio was named in his honour in 1898.

The chemical technology to cure or vulcanize natural rubber was soon employed in newly invented vehicles. The invention of the first pneumatic or inflatable bicycle tyre by the Scottish inventor John Boyd Dunlop, followed by the first internal combustion-powered car of Carl Benz in 1888 with metal wheels covered with rubber, both helped grow the modern tyre industry as the popularity of bicycles and cars grew. The French Michelin brothers pioneered the manufacture of pneumatic tyres to enter the Paris–Bordeaux car race. A pneumatic tyre contained pressurized air as the core and an outer casing made of fabric strengthened with rubberized fibre cords (plys) embedded in the rubber wall and running in a diagonally alternating fashion. The early tyres were 'slick' tyres without treads and had only a thin layer of rubber, making them liable to punctures and slippage between the wheel and the road. As speeds increased, tyres with a tread pattern were introduced, which increased the friction to the road. Apart from tyres, rubber had many uses in cars, including door and window sealing, engine seals and gaskets, hoses, belts, shock absorption and flooring.

In 1913 Henry Ford introduced the first assembly lines for mass car production, which lowered costs, increased car sales and drove greater demand for costly Asian rubber. This led to research work to synthesize artificial rubber polymers by pioneering chemists and technologists who had begun substituting naturally occurring substances with products of the chemical industry – such as synthetic dyes derived from coal tar and celluloid or Bakelite to replace natural ivory or insect-based shellac.

Goodyear manufacturing factory (1908)

Synthetic rubber

Michael Faraday had first determined that rubber had the empirical formula $[C_5H_8]$. The first synthetic rubber was prepared by Gustave Bouchardat in 1879, who heated isoprene liquid with hydrochloric acid to produce a rubber-like polymer. Another synthetic rubber was made by

William Tilden three years later using isoprene obtained from cracking turpentine [$C_{10}H_{16}$], but it was a slow process.

In 1910 the Bayer Company in Germany built a pilot plant to make synthetic rubber by polymerizing 'methyl isoprene' (2,3-dimethylbuta-1,3-diene), which was used during World War I when blockades halted the import of natural rubber. This rubber was more expensive and inferior to the natural material so it was abandoned after the war. The Russians Sergei Lebedev and Ivan Ostromislensky made a rubber polymer from butadiene and pioneered the study of synthetic rubber technologies while working in US rubber companies. Researchers at IG Farben, the German chemical conglomerate, focused on the sodium polymerization of butadiene to produce a synthetic rubber known as *Buna* (from butadiene and natrium for sodium). They discovered that Buna S (butadiene and styrene copolymerized in an emulsion) when compounded with carbon black was significantly more durable than natural rubber. Chemists at the oil company Standard Oil of New Jersey began research and development on the production of butadiene from petroleum on a large scale. When the natural rubber supply from Asia was again threatened in World War II, the US and its allies faced a very serious loss of this strategic material that had become essential for cars and military vehicles. Aircraft, tanks and battleships too all required rubber components, while soldiers needed it for footwear, clothing and a variety of military equipment. It was vital that almost a million tonnes of natural rubber imports was urgently replaced with a synthetic substitute. Allied bombing was directed at the synthetic rubber factories in Germany and Italy. In 1940 B.F. Goodrich Company built a 45 kg per day pilot plant to copolymerize butadiene with methyl methacrylate to produce *Ameripol* rubber for tyres. Goodyear Company in Ohio patented *Chemigum*, another synthetic rubber, while Firestone was also polymerizing butadiene and styrene to produce a synthetic rubber.

In December 1941 the US rubber companies, with government sponsorship, agreed to collaborate in research with universities and the petrochemicals industry. In record time, this wartime spirit of technical co-operation led to new large-scale synthetic rubber production. The German Buna S rubber process was used, as this could be mixed with natural rubber, milled on the same machines and the raw materials were widely available. This rubber was well suited for tyres because it resisted abrasive wear and it retained sharper impressions in moulds than natural rubber. Soon, the first plants were producing rubber because of the massive co-operative effort and impressive chemical engineering achievements. By 1944 a total of 50 factories across the US were manufacturing the GR-S (Government Rubber Styrene) using potassium persulfate as a catalyst in volumes twice that of the world's natural rubber production before the beginning of the war.

After the war natural rubber was again available, but new synthetic rubbers found applications as diverse as solid fuel rocket motors for missiles and space rockets, insulating wiring, printing of textiles and even chewing gum manufacture. Polyisoprene (*cis*-1,4-polyisoprene) was made from the polymerization of synthetic isoprene, which was widely available from oil cracking, to the extent that by the early 1960s production of synthetic rubber exceeded that of natural rubber.

Styrene-butadiene rubber (SBR), with its hard-wearing properties, remains the most prevalent synthetic rubber used in the production of tyres, particularly for cars and vehicles and in conveyor belts, hoses, seals and gaskets. Other rubbers include butyl rubber, which is a copolymer of isobutylene with isoprene that is used in tyre inner tubes or linings owing to its resistance to diffusion of air. Neoprene rubber was developed at DuPont from polymerization of chloroprene (2-chlorobuta-1,3-diene). This rubber is highly resistant to heat and chemicals such as oil and petrol. Silicon-based synthetic rubbers are less sensitive to cracking than other rubbers and have often been used in keypads.

Driving into the future

The chemical technology of vulcanized rubber remains an integral part of modern tyre production. New rubber tyres have been developed for increased strength, reduced weight, improved handling and fuel economy. Many of the tyre innovations have resulted from motor sport. The radial-ply tyre with steel belts offers a softer ride, less rolling resistance, higher fuel efficiency, puncture resistance and a longer life. Other developments such as tubeless and 'run-flat' tyres were created to meet specific needs.

The global demand for tyres is valued at over $250 billion and at least two billion vehicle tyres are made annually, requiring 60% of the world's rubber supply. The demand continues to grow as global incomes rise accompanied by increasing car and vehicle ownership in emerging economies. Nowadays, the tyre industry is a large multinational one. It continues researching further improvements to tyre road performance, but also ways for recycling and finding new materials that are not dependent on fossil fuels to produce tyres for use in the electric and hydrogen vehicles of the future. Synthetic rubber has become a vital part of transportation, aerospace, manufacturing, electronics and myriad consumer products.

U: Ultraviolet and atmospheric chemistry

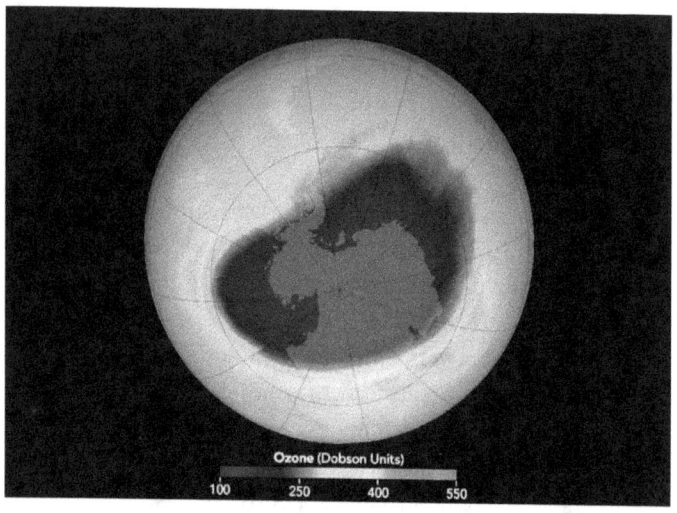

NASA monitoring of Antarctic ozone layer

The hole in the sky

Ultraviolet (UV) light is an invisible form of electromagnetic radiation having a shorter wavelength (10–400 nm) than visible violet light. It constitutes about 10% of the total electromagnetic radiation from the sun. Earth would not be able to sustain life on the land if most of the solar UV radiation were not filtered out by the atmosphere. Most UV from the sun is absorbed by the *ozone* layer. This layer is in the region of earth's stratosphere 20 km to 50 km above the surface. Ozone [O_3], an unstable allotrope of oxygen, is a pale blue gas with a distinctively pungent smell that is formed from oxygen in air by the action of the UV light and atmospheric electrical discharges. People and plants live with both the helpful and harmful effects of UV radiation. Having a low penetrating power, its effects on humans are mainly limited to the skin, where it stimulates production of vitamin D and causes suntan, sunburn or ageing effects. However, in higher exposures it has the potential to cause skin cancer.

The unforeseen interaction of some useful products of chemical technology – the highly beneficial chemicals previously employed in most kitchen fridges, refrigeration systems, aerosol cans and as cleaning agents – with the UV protective ozone layer in the earth's upper atmosphere caused chemists, earth scientists and health experts to realize that, uncontrolled, these products could dramatically affect the earth's atmospheric chemistry.

UV radiation was first discovered in 1801 when the German physicist Johann Wilhelm Ritter observed that invisible rays just beyond the violet end of the visible spectrum darkened silver chloride-soaked paper more rapidly than visible violet light. In 1878 the sterilizing effect of the short-wavelength light for killing bacteria was discovered. It is possible to artificially produce UV radiation in gaseous discharge tubes, specialized lamps and electric arcs. It can be used to treat jaundice in newborn babies, sterilize medical equipment, enhance suntans and produce artificial light effects. UV spectroscopy is widely used as a technique in chemistry to analyse chemical structures.

The UV radiation at the longer wavelengths of 320–400 nm (known as UV-A) plays an essential role in the formation of vitamin D. This vitamin helps regulate the amount of calcium and phosphate in the body and keeps bones, teeth and muscles healthy. Its absence can lead to bone deformities such as rickets in children. UV radiation at shorter wavelengths of 290–320 nm (known as UV-B) causes damage to the skin and impacts the fundamental building block of life – DNA. As a result, distorted proteins are made or cells can die. Over millions of years of evolving in the presence of UV-B radiation, cells have developed the ability to repair DNA, which provides some resilience to damage. In the atmosphere, ozone occurs in very small quantities, but it plays a fundamental part in life on earth. From an environmental standpoint, in the stratosphere (above 20 km) ozone gas usefully absorbs potentially damaging UV radiation, but in the lower troposphere, the region of the atmosphere from earth's surface to 12 km up, ozone is a pollutant and a component of photochemical 'smog'.

In the mid-1970s it was determined that there was a human-generated threat to the protective ozone layer. The cause of the depletion was linked to the broad class of chemicals found in fridges and used in aerosol spray cans known as *chlorofluorocarbons* (CFCs). From the 1870s, when refrigeration technology first became essential in food storage, most of the early refrigerants (liquids that transfer heat from inside the refrigerator to outside) were either highly toxic or highly flammable, such as ammonia or sulfur dioxide. As a result, there were many deaths from poisoning or fires from leaking refrigerants. To prevent these hazards the leading companies, Frigidaire, DuPont and General Motors, collaborated to find a refrigerant that would be much safer to use. A team of chemists led by Thomas Midgely (1889–1944) worked to develop nontoxic, non-flammable alternatives to the traditional refrigerants.

The result was marketed by DuPont as *Freon*, a mixture of CFCs. *Freon* was thought so safe that its inventor once both inhaled it and then breathed it out onto a candle in front of the American Chemical Society in 1930. The CFC refrigerants resulted in safer and cheaper refrigerators being produced for home and commercial use. The price of refrigerators dropped by 40%, allowing ownership of refrigerators to become widespread. Midgely received many awards for his contributions to chemistry. By the early 1970s worldwide production of CFCs had reached nearly one million tonnes per year with sales of $500 million.

Emerging scientific concern over ozone depletion in the upper atmosphere prompted extensive efforts to assess the potential damage to life due to the increased levels of UV radiation passing through the atmosphere to the surface, particularly in southern latitudes during certain times of the year.

Pioneers

It is Sherwood Rowland (1927–2012), a professor of chemistry at the University of California, and Mario J. Molina (born 1943), a postdoctoral fellow in his laboratory, who are considered pioneers in stratospheric ozone research.

They were the first to show that CFCs could destroy the ozone in earth's stratosphere. The stratospheric ozone absorbs UV radiation before it reaches the earth's surface. They found that the inert and essentially non-toxic CFCs were causing a threat to life. Rowland's interest in the fate of CFCs in the atmosphere dated from 1972 when he learned of the work of James Lovelock, the British scientist who in 1957 had invented a highly sensitive way to measure trace gases. Lovelock had measured trichlorofluoromethane (CFC-11) in the atmosphere in amounts that suggested that practically all this gas ever manufactured was still present. Rowland decided to investigate the fate of CFCs in the atmosphere. When CFCs leaked from faulty equipment, gas storages or when disposing of old fridges, they started to reach significant levels in the upper atmosphere. He realized that they can be broken down by UV radiation once they drift up into the stratosphere. When exposed to UV light in the upper atmosphere, the CFC molecules release a chlorine atom, which being highly reactive catalyses the breakdown of ozone [O_3] to molecular oxygen [O_2]. As is the case with all catalysts, the rate of this reaction is increased but they are not consumed in the reaction, so that a single CFC molecule releases a chlorine atom that can react immediately with an ozone molecule, setting off the chain reaction to destroy millions of ozone molecules. This had started to cause the large-scale depletion of the protective ozone layer.

In late 1973, Rowland and Molina calculated that CFC molecules released near the surface of earth would, over decades, wind up in the stratosphere. They estimated that, even if CFC use was banned immediately, the ozone loss would go on for years.

Rowland said: *'When we realized there was a very effective chain reaction, that changed the CFC investigation from an interesting scientific problem to one that had major environmental consequences.'*

In 1974, in a paper in the journal *Nature*, they finally concluded: *'Chlorofluoromethanes are being added to the environment in steadily increasing amounts. These compounds are chemically inert and may remain in the atmosphere for 40 to 150 years, and concentrations can be expected to reach 10 to 30 times present levels. Photodissociation of the chlorofluoromethanes in the stratosphere produces significant amounts of chlorine atoms, and leads to the destruction of atmospheric ozone.'*

Rowland and Molina were jointly awarded the 1995 Nobel Prize in Chemistry for their pioneering work in understanding that some chemical technologies could affect the environment on a global scale.

In 1976 the National Academies of Science issued a report affirming the destructive effects of CFCs on stratospheric ozone, and governments began considering a ban on the use of CFCs in aerosol cans, although the data on CFCs affecting the ozone depletion were inconclusive as the chemistry was difficult to replicate in the laboratory and ozone concentrations fluctuate naturally with geography and season. The crucial evidence came from British scientists working at the Halley Station of the British Antarctic Survey, who had been taking ground-based measurements of total ozone for decades. British geophysicist Joseph C. Farman (1930–2013) and his colleagues studied the data and found that stratospheric ozone had decreased greatly since the 1960s. By 1985 they concluded that stratospheric ozone over Antarctica had reduced by 40%. The discovery of this so-called 'ozone hole' in the Antarctic confirmed depletion of the ozone layer, which had potential to harm people from a significant increase in the incidence of skin cancer.

Cylinder of HFA refrigerant R134A

Global action

CFCs used as refrigerants, in air conditioners and in aerosol cannisters had been changing the chemistry of the atmosphere in a way that reduced ozone in the stratosphere. To prevent this happening required international co-operation by governments worldwide. In 1987 it was agreed to act under the Montreal Protocol, an international treaty that proposed dramatically cutting CFC production and use. Production of CFCs and other ozone-depleting chemicals was to cease. NASA has measured the earth's stratospheric ozone layer by satellite since 1979, showing that, since the protocol went into effect ozone levels have stabilized but complete recovery is still a few decades away.

Although it was possible to phase out the use of CFCs as propellants in aerosol cans, it was more challenging to ban their use in refrigerators and air conditioners without safe alternatives being available. Chemical technologists developed a related group of compounds known as *hydrofluoroalkanes* (HFAs), such as R-134a (1,1,1,2-Tetrafluoroethane) a common refrigerant, but these presented a different drawback. They are potent greenhouse gases but there are very few commercially available alternative refrigerants that are non-toxic and non-flammable while having the appropriate thermodynamic properties. The understanding of the importance of UV radiation protection provided by the ozone layer holds lessons for future chemical technologies when dealing with global climate change in the 21st century.

V: Vaccines

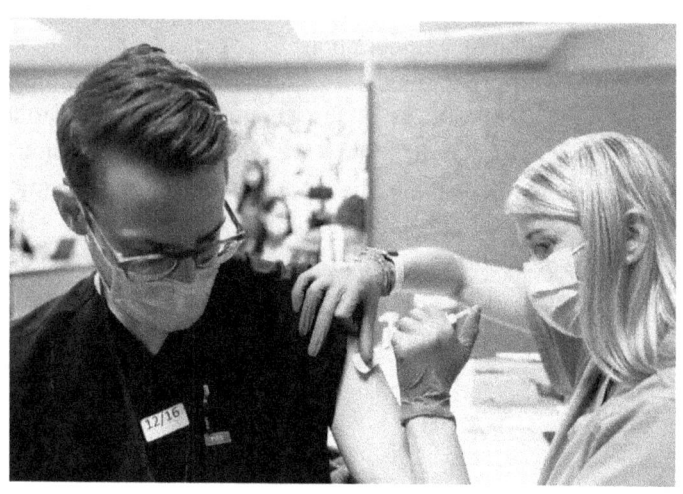

'The tugboats of preventive health'–William Foege

Vaccines have been responsible for eradicating or reducing many serious diseases such as smallpox, diphtheria, polio, tetanus and measles that were once widespread. While involving biological rather than purely chemical technology, the production of modern vaccines has required advances in multiple areas including chemistry, biochemistry, medicine, epidemiology, microbiology and molecular biology. Vaccination is one of the most important ways to protect against ill health, preventing millions of deaths worldwide every year and extending lifespan. The basic strategy behind the use of vaccines is to prepare the human immune system to deal with harmful pathogens.

A vaccine is made by first generating the *antigen*, a substance capable of stimulating an immune response, such as activating lymphocytes, the body's infection-fighting white blood cells. The antigen can take various forms, such as an inactivated virus or bacterium, an isolated portion of the infectious agent or a recombinant protein made from the agent. The antigen is then isolated and purified, and a variety of substances are incorporated to hasten the body's immune response or enhance activity and ensure its stable shelf life. The successful manufacture relies on the skills and technologies of the chemicals and pharmaceutical industries. The first modern vaccine was introduced by British physician Edward Jenner.

Edward Jenner vaccinating a child against smallpox

Pioneers

In the early 18th century smallpox was a serious ancient disease caused by the *variola* virus. It was widespread, very often fatal (1 in 3 died) and greatly feared by all social classes. In 1716, while in Istanbul in the Ottoman Empire, Lady Mary Montagu (1689–1762), an English aristocrat and writer, observed the practice of 'variolation' and promoted the method on her return to Britain. This involved intentionally infecting a healthy person with the dried crusts from lesions taken from a sick patient with a mild attack of the disease in the hope that it would provide immunity from further infection. Unfortunately, the disease did not always remain mild, and the inoculated person could then spread the disease to others. In 1798 Edward Jenner (1747–1823) employed a substance that was safer than the actual smallpox virus to obtain immunity.

After secondary school Jenner had been apprenticed to a surgeon where he acquired a knowledge of medical and surgical practice. Later he was tutored by John Hunter, at St George's Hospital, who was one of the most prominent surgeons and experimentalists in London. Hunter was concerned with the problems of physiology and body function. Jenner also developed an interest using his disciplined powers of observation and a reliance on experimental investigation. He had observed that a person who had suffered an attack of cowpox, the relatively harmless disease that could be contracted from cattle, could not become infected by accidental or intentional exposure to smallpox. In May 1796 Jenner met a young dairymaid, Sarah Nelmes, who had fresh cowpox lesions on her hand. Using samples from her lesions, he inoculated an 8-year-old boy, James Phipps, who had never had smallpox. Phipps became slightly ill over the course of the next week but was well again on the tenth day. Later, Jenner inoculated the boy again, this time with actual smallpox. He saw no disease had developed and so confirmed the protection was complete. His method of vaccination rapidly proved its value and Jenner became active promoting it. By using the related cowpox virus to confer protection he was exploiting a rare situation in which immunity to one virus confers protection against another viral disease. As the death rate from smallpox plunged Jenner received worldwide recognition. In Russia, Empress Catherine the Great became the first to be inoculated to encourage her people. In the 1970s, epidemiologist William Foege (born 1936) working in the Centres for Disease Control and Prevention in the US, devised the strategy of widespread vaccination which finally eradicated smallpox worldwide by 1980.

The next significant vaccine was developed against the deadly rabies virus by the French chemist and biologist Louis Pasteur (1822–1895). His work was to make vaccination an essential part of medicine. Pasteur was born in Dole, France, into a family of leather tanners. He attended the École Normale Supérieure, the famous teachers' college in Paris, earning his master's degree there in 1845 and then a doctorate in 1847.

As a result of his scientific research, he was appointed to the faculty of science in Strasbourg and then in Lille. There he launched his studies on fermentation (see Y). By the 1860s Pasteur and a small number of other scientists believed that diseases arose from the activities of micro-organisms ('germ' theory). He was able to determine that the cause of the devastating blight that had befallen the silkworms in France's silk industry was two separate micro-organisms. He then realized that spoiling of wine was caused by unwanted micro-organisms, which could be destroyed by heating the wine to between 60 °C and 100 °C. His sterilizing *pasteurization* process was later extended to many spoilable substances, notably milk.

In his research on fowl cholera, he discovered how to make vaccines by attenuating, or weakening, the microbes involved. He then realized that the technique could be applied to other diseases such as anthrax, a serious infectious illness of farm animals caused by a soil-based microbe. Pasteur produced a vaccine from weakened *anthrax bacilli* that could protect sheep and other animals from infection. In public demonstrations in mid-May 1881 24 sheep, one goat and six cows were subjected to a course of inoculations with the new vaccine. Meanwhile a control group of stock was left unvaccinated. At the end of the month all the animals were inoculated with virulent *anthrax bacilli*, and two days later the audience reassembled. The effects of the vaccine were undeniable: the vaccinated animals were all alive while the control animals had died or were very ill.

After his success on anthrax, Pasteur then studied the deadly human disease rabies, which while rare is a very serious infection of the brain and nerves. It is caused by several viruses, acquired from the bite or scratch of an infected animal, most often dogs. Rabies is almost always fatal once symptoms appear. The previous treatment for a bite by a rabid animal had been cauterization with a red-hot iron in hope of destroying the unknown cause of the disease, but this was rarely successful. Rabies presented new obstacles to the development of a successful vaccine, primarily because the micro-organism causing the disease could not be specifically identified, nor could it be cultured in the laboratory; but experiments could be done on animals. Attenuation of rabies was first achieved in monkeys and then in rabbits, and Pasteur was able to protect dogs, even those already bitten by a rabid animal. In 1885 he agreed with some reluctance to treat his first human patient, Joseph Meister, a 9-year-old boy who was otherwise facing certain death. The treatment proved to be a pioneering success that was soon known throughout the world.

In 1888 the *Pasteur Institute*, a non-profit private foundation dedicated to the study of biology, micro-organisms, diseases and vaccines, was founded in recognition of his work and it continues today as one of the leading global institutions of biomedical research.

Polio vaccine

An important innovation in vaccine development was made by virologist Jonas Salk (1914–1995) at the University of Pittsburgh, after the US had suffered its worst polio epidemic in 1952 with over 57,000 cases leading to 3,000 deaths. Polio (poliomyelitis) is a highly debilitating, contagious disease caused by a virus that attacks the nervous system, destroying nerve cells in the brain and spinal cord. It can paralyse parts of the body and can affect the muscles used for breathing, so causing a drawn-out death by suffocation. The disease was often contracted by children, and infected patients were often placed in an *iron lung* machine with a cycling pressure to induce inhalation and exhalation. Some patients would be forced to live inside a machine for years. Salk developed a 'dead' or inactivated poliovirus vaccine (IPV), which was given by injection. Approved for use in 1955, it was effective in preventing most of the complications of polio. Soon another researcher, Albert Sabin (1906–1993), working at the Cincinnati Children's Hospital, demonstrated that the poliovirus multiplied and attacked the intestines before it moved to the central nervous system. He developed an oral poliovirus vaccine (OPV) based on mutant strains of polio virus that seemed to stimulate antibody production but not to cause paralysis. It worked in the intestines by blocking the polio virus from entering the bloodstream. It was first licensed in 1961 with the vaccine put on a sugar cube, making it easier to administer, and its effect lasted longer than the original vaccine. The Sabin vaccine became the predominant method of vaccination against polio for the next three decades. It broke the chain of transmission of the virus and opened the possibility that polio might one day be eradicated. Sabin also developed vaccines against other viral diseases, including encephalitis and dengue fever.

Many more have become available recently, such as, a pneumococcal vaccine. Pneumococcus is a bacterium that can cause pneumonia, meningitis and sepsis-it affects young children particularly hard, with some 400,000 children under 5, dying each year from pneumonia. Health experts had long believed that a vaccine targeting this bacterium could save millions of children's lives. But making the first safe and effective pneumococcal vaccine for young children would prove challenging. In 2017 a new cost-effective vaccine was developed (driven by the *Gates Foundation*) for use in poorer countries, allowing over 150 million children to be immunized and saving an estimated 700,000 lives. There are many other diseases for which vaccines have been developed. Vaccines against viruses provide especially important immune protection since, unlike bacterial infections, viral infections do not respond to treatment from antibiotics.

Vaccine types

The challenge is to devise a vaccine strong enough to treat infection without making an individual seriously ill. To that end, researchers have devised different types of vaccines. These include the previously mentioned weakened or *attenuated* vaccines, which consist of micro-organisms that have lost the ability to cause serious illness but retain the ability to stimulate immunity. Attenuated vaccines include those for polio, measles, mumps, rubella and tuberculosis. *Inactivated vaccines* are those that contain organisms that have been killed or inactivated with heat or chemicals. Vaccines against rabies, polio, some forms of influenza and cholera are made from inactivated micro-organisms. Another type of vaccine is a *subunit vaccine*, which is made from proteins found on the surface of infectious agents, such as vaccines for influenza and hepatitis B. The metabolic by-products of infectious organisms may be inactivated to form *toxoid vaccines*. They induce an immune response to the original toxin or increase the response to another antigen since the markers are preserved. Vaccines based on toxoids have been used to provide immunity against tetanus, diphtheria and pertussis (whooping cough).

By the late 20th century, advances in genetic techniques allowed new approaches to vaccine development. Medical researchers could identify the genes of a disease-causing micro-organism that encode the protein or proteins which stimulate the immune response to that organism. This allowed the immunity-stimulating proteins (antigens) to be mass-produced for use in vaccines. It also made it possible to alter pathogens genetically and produce weakened strains of viruses, providing safer and more effective methods by which to manufacture attenuated vaccines. Recombinant DNA technology (see R) has also proven useful in developing vaccines to viruses that cannot be grown successfully or that are inherently dangerous. Genetic material that codes for a desired antigen is inserted into the attenuated form of a large virus, such as the *vaccinia virus*, the source of the modern smallpox vaccine. The altered virus is injected into an individual to stimulate antibody production to the foreign proteins and thus confer immunity. Vaccines against human papillomavirus (HPV), responsible for cervical cancer and other serious diseases, are made from virus-like particles (VLPs), which are prepared using recombinant technology. The vaccines do not contain live virus or genetic material and therefore are incapable of causing infection. Still another approach, called 'naked' DNA therapy, involves injecting DNA that encodes a foreign protein into muscle cells. The cells produce the foreign antigen, which then stimulates an immune response.

Most vaccines contain an active component (*the antigen*) that generates an immune response. They also contain other ingredients to keep the vaccine safe and effective. *Preservatives* are required to prevent the vaccine from becoming contaminated once the vial has been opened. There are *stabilizers* to prevent chemical reactions from occurring within the vaccine and keep the vaccine components from being retained in the vaccine vial. *Surfactants* are used to keep all the ingredients in the vaccine blended and prevent settling in the liquid vaccine. Next, a *diluent* like sterile water is used to dilute a vaccine to the correct concentration prior to use. Finally, some vaccines also contain *adjuvants* to improve the immune response to the vaccine by, for example, stimulating local immune cells using a trace amount of an aluminium salt. Once the vaccine has been made in bulk quantities, it is put in vials and then carefully packaged for safe cold storage and transport to medical facilities.

Many vaccines have been safely used globally over many years, with millions of people receiving them. As with all medicines, every vaccine must go through extensive and rigorous testing to ensure it is safe before it can be given to people. The preclinical phase is done without testing on humans. An experimental vaccine is first tested in animals to evaluate its safety and potential to prevent disease. After a series of clinical trials with patients to determine efficacy and safety, regulatory approval must be obtained. Regulators in each country review the trial results and decide whether to authorize the vaccine for use in an immunization programme. The bar for vaccine safety and efficacy is extremely high, recognizing that vaccines are given to people who are otherwise healthy and free from illness. The whole process, from the preclinical trial to commercial manufacture, could take up to a decade to complete.

As the safety of the vaccine is paramount, once vaccines start being administered, national health authorities and pharmaceutical companies constantly assess how effective the vaccine performs in the population and monitor for any possible adverse side effects and responses from people who have received the vaccine over a long time frame. There are always a small minority who believe that vaccines are unsafe, the science flawed or that they infringe on their human rights but the evidence shows that they are an extremely effective health measure.

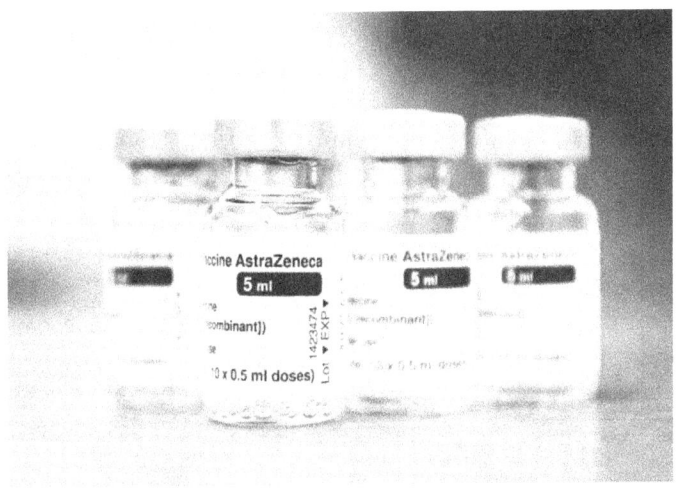

Vials of the AstraZeneca Covid-19 vaccine

Jabs for everyone

Sometimes new approaches are required driven by special circumstances. In response to the global COVID-19 pandemic in 2020, vaccines were urgently developed to provide acquired immunity against severe acute respiratory syndrome coronavirus (SARS-CoV-2). Pharmaceutical companies produced and distributed billions of doses of the new vaccines in response to the worst global health emergency for a century. Fortunately, prior to the pandemic, there was established work on the structure of coronaviruses that had caused diseases like SARS and MERS, which enabled accelerated development of vaccine technologies.

In the urgent search for a COVID-19 vaccine, researchers worked on many different activities in parallel, backed by significant financial and political commitments from governments and pharmaceutical companies. The UK was the first country to deploy an approved COVID-19 vaccine. Development took less than one year – a record time for a vaccine to be available in a mass programme of vaccination that delivered millions of life-saving doses. It was developed jointly by Oxford University and British pharmaceutical company AstraZeneca with funding from the British government. It was a *viral vector* vaccine given by intramuscular injection for preventing symptomatic COVID-19 and illness. This vaccine used the body's own cells to elicit an immune response. There were other vaccines around the world based on a variety of different types.

The technology of vaccines has virtually eliminated many diseases that previously were common in many countries including smallpox, diphtheria, tetanus, poliomyelitis, measles, mumps, rubella, tuberculosis, hepatitis B, meningitis, yellow fever, human papillomavirus and certain types of influenza. There is now the prospect of using a new vaccine (*R21/Matrix-M*) developed by the Jenner Institute at Oxford University for preventing malaria, which still causes over 600,000 deaths globally each year.

W: Water treatment

Building London's waterworks in 1910

**'Water, water, everywhere, Nor any drop to drink' –
Samuel Coleridge**

Modern water treatment was made possible using chemical
technology, which alongside improvements in sanitation
was one of the greatest public health achievements of the
20th century. Clean water is a finite and irreplaceable
resource but has been central in the development of
civilization. Only about 2.5% of water on earth is fresh
water, with over two-thirds of this found in glaciers or
permanent ice sheets. The remainder is salt water contained
in seas and oceans. It was long recognized that poor-quality
water often posed a health risk. Today's safe drinking
(potable) water supplies are largely a result of disinfection

treatment, often by water chlorination, which is the process of adding chlorine or chlorine compounds such as sodium hypochlorite to water to prevent the spread of serious diseases such as cholera, dysentery and typhoid.

History

Early water treatment was primarily focused on the apparent properties of water – taste, clarity and odour. Written works dating back to 2000 BC contain references to methods for water treatment involving heating and using sand and gravel filtration or some means of straining to remove suspended particles prior to boiling. The primary motivation was to improve the taste of drinking water and reduce turbidity, without any knowledge of chemical contamination or micro-organisms. The Egyptians discovered the principle of coagulation around 1500 BC, employing alum salts containing sulfates of iron, potassium and aluminium to cause suspended particles to settle out. In Ancient Greece, processes of boiling and filtering water through charcoal were used in addition to straining. The Greek physician Hippocrates invented a water-sieving process around 500 BC using the 'Hippocratic sleeve', a cloth bag filter to remove sediments that gave water a bad taste or smell. The Romans built extensive aqueducts to transport water over extremely long distances by gravity. Many aqueducts were underground structures, to protect them in times of war and to prevent contamination. After the collapse of the Roman Empire in the 5th century, much of this infrastructure fell into disrepair or was destroyed by invaders in the early Middle Ages.

By the 17th century, as populations grew in Europe, spreading serious diseases and leading to water shortages, work on more effective water treatment began. In 1627 Sir Francis Bacon experimented with seawater desalination, unsuccessfully attempting to remove salt from seawater by passing it through a sand filter. In the 1670s Antonie van Leeuwenhoek and Robert Hooke used the newly invented microscope to observe for the first time the small particles that were suspended in raw water. This began the understanding of waterborne micro-organisms and pathogens. The first water filters made from charcoal, wool and sponge were created for domestic homes in the 1700s, and by 1804 the first municipal water treatment plant, designed by engineer Robert Thom (1774–1847), was built in Scotland. This supplied the town of Paisley including its textile works and its entire population with drinking water. Within three years filtered water was being transported to the larger city of Glasgow. In 1806 Paris also started its first large water treatment plant. The raw water was settled for 12 hours before it was filtered in beds consisting of sand and charcoal, which were replaced every six hours.

The Chelsea Waterworks Company in London had been founded in 1723 to supply water to central London, but as demand grew there were increasing concerns about the quality of the water being drawn from the River Thames. In 1829 the civil engineer James Simpson (1799–1869) built a large-scale sand filtration system that provided filtered water for London. The principle was widely copied throughout the UK. As pollution of the Thames increased, an Act of Parliament was passed in 1852 to prohibit the extraction of water for household purposes from the river in its more polluted reaches below the Teddington Lock. Unsanitary conditions continued to be a feature of mid-19th-century London. After a severe cholera outbreak in 1854 it was discovered that cholera bacteria had spread through water, but it was apparent that the outbreak seemed less severe in areas where sand filters were installed. A

British physician, John Snow, was famously to find that the cause of the outbreak was a communal water pump that had become contaminated by raw sewage.

Pioneer

John Snow (1813–1858) was born at York in England. His father was a labourer in a local coal yard and then a farmer to the north of the city. As a child, Snow was living in an area that was frequently in danger of flooding because of its proximity to the river and experienced unsanitary conditions due to sewage-contaminated water. He demonstrated an aptitude for mathematics at school and he obtained a medical apprenticeship in Newcastle-upon-Tyne. In 1832 he encountered a cholera epidemic for the first time in Killingworth, a coal-mining village, while treating many of the victims. In 1836 he enrolled at the Hunterian school of medicine in London, then began working at the Westminster Hospital. He graduated from the University of London in December 1844 and set up his practice in Soho as a surgeon and general practitioner. His first interest was anaesthesiology. He was one of the first physicians to study and calculate dosages for the use of ether and chloroform as surgical anaesthetics, allowing patients to undergo surgical and obstetric procedures without the distress and pain they would otherwise experience. In 1853 Queen Victoria asked Snow to administer chloroform for pain relief during the delivery of her eighth child, Prince Leopold. This led to wider public acceptance of its use for pain relief in childbirth.

Snow was sceptical of the 'miasma theory' that believed diseases were caused by noxious 'bad airs'. In 1855 he demonstrated the role of the water supply in spreading the cholera epidemic in Soho, with the use of a distribution map, providing statistical proof to illustrate the connection between the quality of the water source and pattern of cholera cases. His data convinced the local authorities to disable the water pump, which soon ended the outbreak. It was discovered that a public well had been dug less than a metre from an old cesspit, from which had leaked the harmful bacteria. Snow's study proved to be a major landmark in the history of public health and is regarded as the foundation of the science of epidemiology – the study of the distribution and causes of disease.

John Snow wrote to the editor of the *Medical Times and Gazette* about his findings:

> *I found that nearly all the deaths (61 in total) had taken place within a short distance of the Broad Street pump. There were only ten deaths in houses situated decidedly nearer to another street-pump. In five of these cases the families of the deceased persons informed me that they always went to the pump in Broad Street, as they preferred the water to that of the pumps which were nearer. In three other cases, the deceased were children who went to school near the same pump ... The result of the inquiry, then, is, that there has been no outbreak or prevalence of cholera in this part of London except among the persons who were in the habit of drinking the water of the above-mentioned pump well ... In consequence of what I said, the handle of the pump was removed on the following day.*

It was Louis Pasteur who was one of the chief proponents of the new 'germ theory' that showed conclusively how micro-organisms could grow and spread in water without careful decontamination and sterilization. Diseases such as cholera, dysentery and typhoid, which had spread through drinking water contaminated by sewers, could be reduced by improved water treatment and sewage arrangements. Snow's findings inspired the fundamental changes in the management of water and waste systems of other towns and cities, leading to a significant improvement in public health around the world. In London, minimum standards of water quality were imposed requiring that all water be filtered from 1856. This was followed up with legislation for the mandatory inspection of water quality, including comprehensive chemical analyses. To ensure the highest-quality intake, automatic pressure filters, where the water is forced under pressure through the filtration system, were used from 1899 throughout Britain. In the US, rapid sand filtration was applied, and subsequently it was found that this worked much better when it was preceded by coagulation and sedimentation treatments.

In Britain, as a legacy of its Victorian designed water and sewer network, the sewage and surface rainwater often flow into the very same pipelines. The problem with a combined system is that when it rains heavily the rain overwhelms the sewage system and a portion of the untreated overflow spills through an outlet into a river or the sea. These spills were supposed to be infrequent, but can cause potential environmental concerns which are expensive to mitigate.

Jardine Water Purification Plant in Chicago

Chlorination

John Snow had also successfully used chlorine to disinfect the water supply in Soho that had helped spread cholera. Chlorine was shown to be effective at purifying water, when smell, taste and appearance alone were previously used to judge safety. Water chlorination is the process of adding chlorine or chlorine compounds such as sodium hypochlorite to water to kill bacteria, viruses and other microbes to prevent the spread of waterborne diseases. Chlorinated lime (bleaching powder) was used to treat the sewage produced by typhoid patients in 1879. Early attempts at implementing a water chlorination process were made in 1893 at a water treatment plant in Hamburg, Germany. In 1897 Maidstone in England was the first town to have its entire water supply treated with chlorine. When a contaminated water supply caused a serious typhoid fever epidemic in Lincoln, England, in 1905, Sir Alexander Houston (1865–1933), an expert from the London Metropolitan Water Board, was sent to investigate. He fed a concentrated solution of chlorinated lime to the water supply. This was not calcium chloride but contained chlorine gas dissolved in lime-water (dilute calcium hydroxide) to form calcium hypochlorite. The chlorination helped stop the epidemic and it was continued until 1911 when a new water supply was commissioned. The first continuous use of chlorine in the US was in 1908 at Boonton Reservoir, which provided the water supply for Jersey City, New Jersey. Over the next few years, chlorine disinfection using calcium hypochlorite was installed in drinking-water systems around the world.

The use of compressed liquefied chlorine gas for the purification of drinking water was separately developed by Vincent Nesfield, a British officer in the Indian Medical Service, in 1903. The chlorine gas was stored in lead-lined iron vessels then bubbled into the water. As a strong oxidizing agent, it oxidized the organic pathogens in the raw water supply from reservoirs and rivers to render it safe. When dissolved in water, chlorine converts to an equilibrium mixture of chlorine, hypochlorous acid [HOCl] and hydrochloric acid. It was first used on a continuous basis to disinfect the water supply at the Belmont plant, Philadelphia, Pennsylvania, employing a 'chlorinator' machine invented by Charles Frederick Wallace in 1913. Nearby in Maryland, Abel Wolman (1892–1989), a pioneering sanitary engineer, helped develop a formula for chlorinating water by considering factors such as acidity, bacteria levels and purity, while still allowing for safe absorption of the dangerous chemical. By 1930 typhoid cases in Maryland had dropped by over 90%. In more developed countries, as a result of the widespread improvements in water treatment, serious waterborne diseases reduced dramatically.

Sometimes chorine disinfection could aggravate respiratory disease or cause an unpleasant taste in the water, and chlorine could react with naturally occurring organic compounds found in water to produce compounds known as disinfection by-products (DBPs). In high amounts, these by-products can cause serious health concerns, so modern drinking water is regularly monitored; however, these risks are low in comparison with the risks associated with inadequate disinfection. Water chemists also developed alternative disinfectants and treatments such as using ozone gas, which was first done in 1906. Serious problems do sometimes still arise, such as the notorious incident in Flint, Michigan, in 2014 where local authority cost-cutting measures led to tainted drinking water containing toxins. The town had been the home of the nation's largest General Motors car plant but it declined from the 1980s. While a new pipeline from Lake Huron was under construction, the Flint River was used as a temporary water source to keep costs down. Soon, residents noticed changes to the drinking water's colour, smell and taste. Analytical testing undertaken found that dangerous levels of poisonous lead dissolved in the river water was reaching residents' homes.

As illustrated by this case, public health concerns in modern times have switched to focus on water contamination from pollutants such as industrial wastes, pesticides and farming residues, requiring water treatment plants to apply a range of new techniques. Modern water treatments often involve additions to the traditional filtration and chlorination technology. These include using activated carbon adsorption or membranes in reverse osmosis (RO) systems. Partially permeable membranes remove many types of dissolved and suspended chemicals and can be employed to produce fresh water from salt water. New techniques such as ultraviolet irradiation, advanced oxidation with ozone, ion exchange and biological filtration are increasingly employed to treat water.

Drinking the benefits

It is because of chemical technologies used in water purification that millions of people worldwide have gained access to safe drinking water. This was one of the greatest public health achievements of the 20th century, with the linked improvements in sanitation and hygiene. As countries develop and populations grow, they industrialize and intensify agriculture, so demand for water increases. Over 70% of global fresh-water resources are used in food production. Global demand for water is expected to exceed supply by 40% in the next two decades making fresh-water security a critical issue.

The United Nations considers that water shortages can pose a serious challenge to sustainable development and may cause conflicts. In response to these concerns, water technologists are creating new ways to manage water, such as economic desalination of sea water, the purification and recycling of wastewater and leak reduction. As they did in the past, chemical technologies can make a major contribution to water management. Developing new technologies will remain vital for reducing the global burden of disease and improving the health, welfare and the productivity of populations, particularly in the light of environmental pressures, such as climate change, and the need to recover valuable materials from wastewater, such as rare metals, nitrates, phosphates and biogas.

X: Xenon

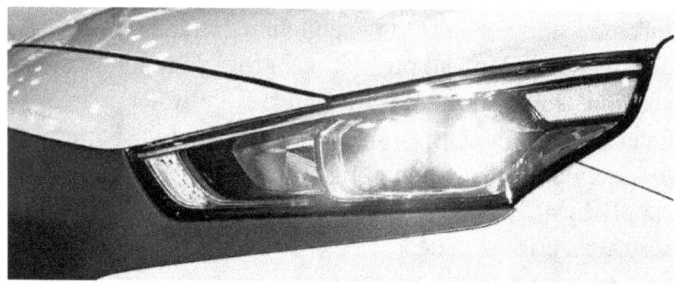

Xenon gas filled headlights

The noble gases

Xenon is a rare gas that has some very useful industrial applications such as in car headlights and lasers. It is used in medicine as an anaesthetic and for imaging the body, in physics research and as a propellant for ion propulsion in spacecraft. The gas was first discovered because of new low-temperature (or cryogenic) technology to liquefy air that was developed by the German engineer Carl von Linde and independently by British inventor William Hampson.

Xenon is a colourless and odourless chemical element (atomic number 54), about 4.5 denser than air and only found in earth's atmosphere in trace amounts (0.09 parts per million). It is a *noble* gas; these gaseous elements are very unreactive because their outer valence shell contains eight tightly bound electrons resulting in a stable, minimum-energy configuration. When a xenon gas-filled tube is subject to an electrical discharge it glows blue. It is used in high-intensity discharge (HID) vehicle headlights and photographic lights to produce much brighter light than halogen bulbs. Xenon was discovered by the Scottish chemist Sir William Ramsay (1852–1916) in 1898. He found the xenon gas in the residue left over from evaporating components of liquid air. He named it xenon from the Greek *xenos*, meaning 'guest'.

William Ramsay went on to discover three other new noble gaseous elements and showed that with helium and radon they formed an entire family of new elements. In addition to xenon, the other noble elements, located in the last column (18th group) of the periodic table, with filled outer shells are helium, argon, krypton, radon and oganesson. At first, the discovery of these noble gases was treated with some doubt by the pioneering Russian theoretical chemist Dmitrii Mendeleev, who had devised the periodic table, as he had not foreseen the possibility of such a group of elements. For a long time, xenon and the other noble gases were considered chemically inert, being unable to form chemical compounds. However, in 1962 Neil Bartlett at the University of British Columbia discovered that platinum hexafluoride could oxidize xenon, allowing the first known compound of a noble gas, xenon hexafluoroplatinate, to be made.

Pioneers

It was the work of two inventors who created special equipment that was to lead to William Ramsay's discovery of xenon and the other noble gases. Since the 18th century only air and a few other gases had been liquefied using time-consuming, expensive and laboratory-scale methods. German Carl von Linde (1842–1934) and William

Hampson (1854–1926) from Britain, who were working independently of each other, went on to create the modern industrial gas industry that supplies vital pure gases such as oxygen, nitrogen and helium.

In the 19th century there was no means for liquefying gases on an industrial scale. As a mechanical engineer, von Linde was producing refrigerators for breweries, abattoirs and cold storage facilities. In 1892 an Irish brewery asked his company to develop a machine that would produce liquid carbon dioxide for beer production. He envisioned that the air itself could be used as a refrigerant in a new liquefaction process by exploiting the 'Joule-Thomson' effect, where as a gas expands it gets colder. Von Linde built an apparatus where a batch of compressed air, pre-cooled to −30°C, would enter a heat exchanger that consisted of a 100-metre-long double steel tube, wound to a spiral, insulated with wool within a wooden case. The effect reduced the temperature of the gas further, cooling the next batch of air even further than the previous one. The cooling kept increasing with every air intake until, at about −190°C, the gaseous air turned to liquid.

On 29 May 1895 von Linde described the moment liquid air first dripped from the plant: *'With clouds rising all around it, the pretty bluish liquid was poured into a large metal bucket. The hourly yield was about three litres. For the first time on such a scale air had been liquefied, and using tools of amazing simplicity compared to what had been used before.'*

Although von Linde registered his patent on 5th June 1895, William Hampson an amateur inventor and former barrister, had already registered a similar process in Britain only two weeks beforehand. Carl von Linde's air liquefaction apparatus took the coveted Grand Prix at the World Fair in

Paris. Both machines relied on the same principles and process, which is now referred to as the 'Hampson–Linde' cycle. In 1893 a Scottish chemist James Dewar had invented a double-walled vacuum storage vessel for liquid gases made of mirrored glass, which was named the Dewar flask. Dewar did not apply for a patent and lost out to the *Thermos* company that first marketed it. Dewar flasks became invaluable for storing and transporting liquified gases and are still used to this day. Von Linde saw that it would be commercially more profitable to be separating air to produce pure oxygen and nitrogen, so he built an air separation plant by adding a rectification step and allowing the separated oxygen vapour to rise slowly up the same column where liquid air was slowly returning down. Von Linde's first oxygen production plant went into operation near Munich in 1903, followed by the first nitrogen plant in 1908 and the first combined air separation plant two years later. At this time, Fritz Haber was working to fix nitrogen from air, which led to synthetic fertilizers (see N) and a huge demand for nitrogen gas production.

Sir William Ramsay, the professor of general chemistry at University College London until his retirement, was able to use the newly designed equipment to discover new gases such as xenon. He was awarded the Nobel Prize in Chemistry in 1904 *'in recognition of his services in the discovery of the inert gaseous elements in air, and his determination of their place in the periodic system'.*

Oxygen and nitrogen gas separation plant, Berlin (1904)

Valuable gases

Carl von Linde's company grew into a major global company, which was to dominate the industrial and specialist gas market. Xenon is still produced commercially as a by-product of the separation of air into oxygen and nitrogen. This separation is followed by cryogenic fractional distillation in a double-column plant, with the liquid oxygen produced containing small quantities of krypton and xenon. These are separated by further distillation. Because of its scarcity, xenon is much more expensive than the lighter noble gases but has several important modern industrial applications. Xenon dissolves in blood and is one of the select group of substances that penetrate the blood–brain barrier, making it of use as a medical anaesthetic. Xenon finds application in medical imaging techniques, such as hyperpolarized magnetic resonance imaging (MRI) to reveal the structure and function of the heart and lungs. It is used in particle physics research and as a propellant for ion propulsion in spacecraft and satellites engines. The first solid-state laser was powered by a xenon flash lamp in 1960. The gases are used in excimer (excited complex) lasers to produce ultraviolet light for high-precision imaging, for microlithography and microfabrication, which are essential for integrated circuit manufacture, for laser angioplasty (the procedure to widen narrowed arteries) and in eye surgery. The global market for xenon alone now exceeds £200 million in value.

All the noble gases are inflammable, colourless, odourless and largely inert in nature. Most have a range of specialized applications in the semiconductor industry, laser technologies, medicine, energy-efficient lighting and aerospace. Commercially helium has many uses, with most available helium being extracted from a small number of natural gas reservoirs in small concentrations (<0.5%) and is a valuable by-product of other processes. Helium being extremely low density is often used in balloons and as a carrier for gas chromatography for chemical analysis. Having a very low boiling point it is used as a super-coolant for cryogenic applications such as MRI, nuclear magnetic resonance (NMR) spectroscopy and in superconducting magnets in particle accelerators. Computer hard drives are filled with helium, being one-seventh the density of air, to reduce power dissipated in air drag and so be more energy efficient.

All noble gases glow in distinctive colours in gas-discharge lamps; these are sometimes called 'neon lights' but they often contain other gases and phosphors, which add various hues to neon's orange-red colour. Argon is also used in lasers and provides an inert atmosphere when welding or preserving food. Radon is an odourless radioactive noble gas released from the normal decay of the uranium, thorium and radium in rocks and soil. In a few areas, depending on local geology, radon dissolves into ground water and without adequate ventilation can accumulate in buildings to levels harmful to health causing lung cancer. The last of the noble gas elements, oganesson, is a purely synthetic element (first made in 2002), which has the highest atomic number (118) of any element in the periodic table. The global annual sales of noble gases combined are over $7 billion.

Y: Yeast fermentation

Dista Products fermentation plant in Liverpool (1981)

'Nature makes penicillin; I just found it' –
Alexander Fleming

Yeasts are single-celled micro-organisms that are fungi. Widely occurring in the natural environment, they have been exploited in fermentation processes for thousands of years to produce beer, wine, bread, soy sauce, cheese and yogurt. One of the most widely commercially used yeast species is *Saccharomyces cerevisiae*, commonly referred to as brewer's or baker's yeast. The biological process of fermentation has been combined with chemical technologies to create many new routes for making important medical and chemical products.

Fermentation is an energy-generating chemical reaction of an organic feedstock by the action of enzymes that are produced by micro-organisms such as yeasts, mould or bacteria. The enzymes are natural catalysts that act using a hydrolysis process to break down complex molecules to form smaller compounds and nutrients. The commercial products of fermentation and related chemical technologies are used not only in foods and brewing but for making pharmaceuticals, famously penicillin, and a variety of industrial chemicals. Moving from its roots in fermentation and employing the skills of chemical technology, the new biotechnology industry began to grow rapidly from the 1940s.

The word 'fermentation' (from Latin *fervere*: to boil) reflected the bubbling and foaming seen in fermenting liquors, resembling boiling. The French chemist and microbiologist Louis Pasteur used the term fermentation for the chemical process by which molecules such as glucose are broken down by yeasts and other micro-organisms grown in the absence of air (anaerobically). In 1857 he was convinced the process was not (as once thought) a magical 'spontaneous generation' but a biochemical activity. Pasteur also recognized that ethanol and carbon dioxide are not the only products of fermentation but also a diverse range of chemicals.

From the greater understanding of fermentation came the industrial application of biology to produce new chemical products. In 1919 Hungarian Károly Ereky coined the term 'biotechnology' to describe the fermentation technology based on converting natural raw materials into more useful products. Modern biotechnology began to develop from its origins in brewing through World War I with a need for larger-scale industrial fermentations to urgently support food production. In Germany, yeast was grown on an immense scale during the war to produce 60% of the country's animal feed. Compounds of lactic acid, a fermentation product, were used to supplement the lack of glycerol, which was needed as a hydraulic fluid. The biotechnology industry in the UK began by feeding the bacterium *Clostridium acetobutylicum* on starch from potato and maize (corn) to produce a mixture of propanone (acetone, which was then in high demand to produce cordite explosive for munitions), butanol and ethanol.

In the 1940s, the most important product of biotechnology became the life-saving antibiotic penicillin. This was first produced industrially in the US using a deep fermentation process originally developed by Pfizer. The company had previous industrial experience with fermenting sugars to produce citric acid. Industrial fermentations would come to be used to make many speciality chemicals in addition to penicillin and citric acid, including dimethyl ketone, vitamins B-2 (riboflavin) and B-12 (cobalamin), protein products, lactic acid and gluconic acid. The production of biofuels such as bioethanol and biodiesel, the generation of feedstocks such as synthesis gas (carbon monoxide and hydrogen) and the production of biodegradable polymers such as the polyhydroxyalkanoates (PHA) are increasingly important. New discoveries on the nature of DNA and genetics have come to provide the tools for modern biotechnology.

Understanding fermentation

The process of fermentation is thought to have been the primary means of energy production in earlier organisms before oxygen was at high concentration in the atmosphere. It was an ancient form of energy production in cells prior to the evolution of aerobic respiration. The reactions of fermentation vary according to the nutrient molecule and end-product involved. In the simple case of using the sugar glucose [$C_6H_{12}O_6$], the end-product is ethanol [C_2H_5OH] and carbon dioxide is released.

Building on Pasteur's work, Eduard Buchner, a German chemist, determined that fermentation was caused by a yeast secretion that he termed *zymase*. In chemical terms, fermentation is an enzyme-catalysed, energy-generating process in which organic compounds act as both donors and acceptors of electrons. In the 1920s, scientists discovered that, in the absence of air, extracts of muscle cells catalyse the formation of lactate from glucose, which indicated that fermentation reactions are not peculiar to the action of yeasts but also occur in many other instances of glucose utilization such as in animal physiology. Glycolysis is an important type of fermentation that is common to muscle cells, yeast, some bacteria and plants. The six-carbon sugar of glucose is oxidized to two molecules of pyruvic acid [$C_3H_4O_3$] yielding a small net gain of chemical energy (ATP) to power living cells. In the absence of oxygen, there are two main pathways for the pyruvate [$C_3H_3O_3$-] end-product. Ethanol fermentation performed by yeast and some types of bacteria breaks the glucose down into ethanol and carbon dioxide, while lactic acid fermentation breaks down the pyruvate into lactic acid. For example, in the production of yogurt, bacteria convert lactose into lactic acid, giving yogurt its sour taste. In vertebrates, during periods of intense exercise, cellular respiration may deplete oxygen in the muscles faster than it can be replenished. The shift to glycolysis produces lactic acid, which may cause muscle cramps and aches.

It was the development of the antibiotic medicine penicillin that transformed the importance of yeast-based fermentation for healthcare. Antibiotics (Greek *anti*: 'against' and *biōtikos*: 'fit for life') have often been called the single most important therapeutic discovery in the history of medicine and referred to as 'miracle drugs'. Antibiotics are naturally produced by various species of micro-organisms to suppress the growth of others and eventually may destroy them.

Natural antibiotics have been employed since early times. For example, a poultice of mouldy bread was applied topically to wounds with beneficial effect dating back to Ancient Egypt. In England, John Parkinson (1567–1650), the apothecary to King James I, was the first to explicitly record the use of mould to treat infections. The first modern antibiotic was arsphenamine (Salvarsan) found by Paul Ehrlich in 1908, who had observed that bacteria took up toxic dyes that human cells did not and sulfa drugs were derived from azo dyestuffs (see M).

The discovery of the most famous antibiotic occurred in 1928. Scottish scientist Alexander Fleming, working in a medical School in London, heralded the dawn of a new 'antibiotic age'. Previously there was no effective treatment for infections such as pneumonia, gonorrhoea or rheumatic fever. Hospitals often saw many people with severe sepsis (blood poisoning) contracted from a simple cut or scratch, but doctors had no real treatments to employ.

Alexander Fleming holding a petri dish (1928)

Pioneer

Alexander Fleming (1881–1955) was born into a farming family Darvel, in Ayrshire, Scotland. He attended the Royal Polytechnic Institution in London and then in 1903 enrolled at St Mary's Hospital Medical School in Paddington, where he qualified with distinction in 1906. He joined the research department at St Mary's, where he became assistant bacteriologist to Sir Almroth Wright, a pioneer in vaccine therapy and immunology. Fleming gained his Bachelor of Science degree with gold medal in bacteriology and became a lecturer there.

Fleming served as an officer in the Royal Army Medical Corps throughout World War I and was mentioned in dispatches. He and many of his colleagues worked in battlefield hospitals on the western front in France, where he witnessed the death of many soldiers from sepsis because of infected wounds. Antiseptics were used at the time to treat infected wounds, but he observed they often worsened the injuries. Antiseptics worked well on the surface, but deep wounds tended to shelter anaerobic bacteria from the antiseptic. His supervisor strongly supported Fleming's findings, but despite this, most physicians over the course of the war continued to use antiseptics even in cases where this worsened the condition of the patients. After the war he returned to St Mary's Hospital, eventually becoming Professor of Bacteriology at the University of London.

It was on a famous occasion in September 1928 that Fleming, returning from a holiday, began to sort through some petri dishes containing colonies of staphylococcus, the bacterium that causes boils, sore throats and abscesses. He noticed something unusual on one dish. It was dotted with bacteria colonies, apart from one area where a mould was growing that was preventing the growth of bacteria. The zone immediately around the mould – later identified as a rare strain of *Penicillium notatum* (since renamed *Penicillium chrysogenum*) – was clear, as if the mould had inhibited bacterial growth. 'Penicillium' is derived from the Latin for 'a painter's brush' as the fronds of this fungus look like an artist's brush.

Fleming found that his 'mould juice' could kill a wide range of harmful bacteria, such as *streptococcus, meningococcus* and the *diphtheria bacillus*. He then set his assistants the difficult task of isolating pure penicillin from the mould. It proved to be very unstable, and they were only able to prepare solutions of crude material for use in his laboratory work. Fleming published his findings in the *British Journal of Experimental Pathology* in June 1929, making a passing reference to penicillin's potential therapeutic benefits. It looked then as if its main application would be in isolating penicillin-insensitive bacteria from penicillin-sensitive bacteria in a mixed culture.

Making penicillin

While Fleming had discovered the therapeutic potential of penicillin, it was Howard Florey at Oxford University who in 1939 isolated penicillin and demonstrated its ability to kill infectious bacteria. It found a major role during World War II, making the life-saving fungus into a widely available medicine to effectively treat bacterial infections, many of which if left untreated would have been deadly. Florey and his colleague Ernst Chain saw how to turn penicillin from a laboratory discovery into a life-saving drug. For their work on penicillin, he needed to process up to 500 litres a week of mould filtrate. His team were forced to grow it in an array of culture vessels such as baths, bedpans, milk churns and food tins before a customized fermentation vessel was designed.

In 1940 Florey carried out vital experiments that showed penicillin could protect mice against infection from deadly *streptococci*. In February 1941 a 43-year-old policeman, Albert Alexander, became the first recipient of the new treatment. He had scratched the side of his mouth while pruning roses and had developed a life-threatening infection, with huge abscesses affecting his eyes, face and lungs. Penicillin was injected and within days he made a remarkable recovery. Unfortunately, supplies of the drug ran out and he died a few days later, but soon other patients began to benefit from the live-saving penicillin. Wartime conditions in Britain made industrial production of penicillin difficult as local companies lacked the capacity for large-scale production and the UK government could not afford to fund the work. Florey travelled to the US to persuade the American pharmaceutical industry to produce penicillin on a large scale. He contacted US experts on penicillin and the leading research laboratory in Peoria, Illinois, with expertise in fermentation. Within a few weeks it was found that it was possible to significantly increase the yield of penicillin by substituting lactose for the sucrose in the culture medium, and then came the even more important discovery that the addition of corn-steep liquor to the fermentation medium produced a ten-fold increase in yield. Later, the Peoria laboratory increased the yield of penicillin further by the addition of penicillin precursors, such as phenylacetic acid, to the fermentation medium.

It was also recognized that Oxford's method of growing the mould on the surface of a nutrient medium was inefficient, and that growth in a submerged culture fermentation would be superior where the mould is grown in large tanks in a constantly agitated and aerated mixture. Florey's original *Penicillium* culture, however, produced only traces of penicillin when grown in a submerged culture, but staff in Peoria screened various new strains and found the most productive one came from a mouldy cantaloupe, an orange-fleshed melon bought in a local fruit market. An even more productive mutant of the cantaloupe strain was produced with the use of X-rays at the Carnegie Institution. When this strain was exposed to ultraviolet radiation at the University of Wisconsin, its productivity was increased still further. US pharmaceutical company Pfizer was asked by the American and British governments to carry out large-scale penicillin production.

Pfizer's director of research John Kane, a chemist asked John McKeen, a chemical engineer and plant superintendent, to develop a pilot plant. In the end, it was these two who made it possible for the world to benefit from this great medicine with advances in chemical technology. The steps of fermentation, recovery and purification and packaging were developed quickly due to the co-operative efforts of the scientists and engineers working on the project. Many pharmaceutical and chemical companies contributed to solving the problems of scaling up submerged fermentation from the pilot to full-scale manufacturing. As the scale of production increased, they faced new engineering challenges. The Pfizer factory manager captured the complexity and uncertainty facing these companies during the scale-up process: *'The mould is as temperamental as an opera singer, the yields are low, the isolation is difficult, the extraction is murder, the purification invites disaster, and the assay is unsatisfactory.'*

Deep-tank fermentation had been known for a long time, having been employed in breweries, but because penicillin needs air to grow, aerating the corn-steep culture medium in deep tanks required the severe foaming to be reduced by adding an anti-foaming agent. During World War I, the shortage of imported lemons and limes to produce citric acid for soft drinks had forced Pfizer's chemical engineers to produce the acid from fermenting strains of *Aspergillis niger* (a common 'black' mould on certain fruits) in sugar solution. The process required deep tanks but again aeration was a problem until a method had been found of bubbling air through the mixture while agitating it with an electric stirrer. The company had acquired skills in deep-tank fermentation and had applied them to a variety of products like fumaric acid, gluconic acid, itaconic acid and vitamins such as riboflavin(B2). Submerged fermentation also required the design of new cooling systems for the tanks and new mixing technology to stir the penicillin liquor (mash) efficiently. Once the fermentation was complete, recovery was also difficult; they often lost two-thirds of the penicillin during purification because of its instability and heat sensitivity. Extraction required low temperatures. It was found that freeze-drying under vacuum gave the best results in purifying the penicillin to a stable, sterile and usable final form.

Pfizer purchased an old ice factory near Pfizer's Brooklyn premises for carrying out the fermentation of penicillin. A spirit of wartime patriotism drove the work, as every worker was told that *'penicillin produced today will be saving the life of someone in a few days or curing the disease of someone now incapacitated'*. McKeen delivered, and he had the Brooklyn plant running within four months. Fourteen 30,000-litre tanks were started in March 1943, and within four months Pfizer produced five times as much penicillin as it had initially forecast. By the end of the year, the company had produced over 45 million units of penicillin, enough for every Allied soldier to be carrying a dose of the antibiotic for D-Day in June 1944. This achievement saved many lives and it established the value of penicillin in the treatment of surgical and wound infections. The defining success of penicillin was not the initial discovery or laboratory work but the technological breakthrough of using deep fermentation for large-scale production.

News of the new 'miracle drug' began to reach the public and doctors, so demand for penicillin increased. Initially supplies were limited as priority was given to the military. Meanwhile, clinical studies proved penicillin to be effective in the treatment of a wide variety of infections. In Britain, the structure of the penicillin molecule, aided by the X-ray crystallographic work, was established in 1945. It was a four-membered β-lactam ring, fused to a thiazolidine ring. In the same year, Alexander Fleming, Howard Florey and Ernst Chain were awarded the Nobel Prize for their penicillin research.

Penicillin production began to increase dramatically to more than 6,800 billion units in 1945 using larger 45,000-litre tanks with 80–90% yields. Soon penicillin was available to the consumer in chemists everywhere. By 1949 the annual production of penicillin in the US was 130,000 billion units and the price had dropped 100%. In the UK, penicillin was first available as prescription drug in June 1946.

Significant improvements in modern production plant and biology have increased production rates and decreased costs. Penicillin was the first β-lactam antibiotic discovered. The penicillin inhibits the bacterial activity by inactivating the penicillin-binding proteins inside the cell wall and finally destroys the cell wall synthesis of the bacteria. The β-lactam antibiotics are the most widely used group of antibiotics, making up more than half of all commercially available antibiotics. Worldwide sales of penicillin and other derivative antibiotics are greater than $10 billion per year, while costing $10 per kilogram. Today, penicillin is used for treating bacterial infections of the skin, respiratory tract, ear, nose and throat, as well as pneumonia, rheumatic fever and those serious infections that might result during surgery.

Most modern antibiotics follow a similar process for their commercial production, with differences in micro-organisms, fermentation media and methods of extraction. Antibiotics are used to treat or prevent some types of bacterial infections but are ineffective against viruses such as the common cold or influenza. Although they have delivered great benefits, the overuse of antibiotics means they have become less effective against serious infections. It is the efficacy, availability and ease of use of antibiotics that have contributed to their overuse, and some bacteria have developed resistance to infections. In 2020 more than 1.2 million people died worldwide from infections caused by bacteria resistant to antibiotics, and this is now a major concern in the therapy of both humans and animals. The overuse of antibiotics has led to the emergence of 'superbugs'. These are strains of bacteria that have developed resistance to many different types of antibiotics, such as MRSA or C. diff. These types of infections are serious and are proving challenging to treat. Using antibiotics for preventative (prophylactic) treatment in the production of chickens, pigs and cattle has also been associated with the emergence of antibiotic-resistant strains of bacteria, including *salmonella*, *campylobacter* and *E. coli*. As a result of these concerns about growing antibiotic resistance, health organizations across the world are attempting to reduce their use for less serious conditions.

Plant producing ethanol from corn, Colorado

Chemicals from biotechnology

Manufacturing with biological processes, using enzymes or microbes, can be successfully employed as a complement to chemical manufacturing with reduced costs and environmental emissions compared with traditional methods. Some chemicals, such as citric acid (2-hydroxypropane-1,2,3-tricarboxylic acid), have been routinely produced on a large scale using biotechnology because the chemical synthetic routes were complex and expensive. Propanediol is widely used in products for skin care, cleansers and liquid detergents, in industrial applications such as anti-freeze and as a monomer to produce unsaturated polyester resins. A manufacturing facility to produce Bio-PDO (1,3-propanediol) from renewable resources using a fermentation process was developed by DuPont and Tate & Lyle using corn sugar instead of petroleum-based feedstocks. This propanediol consumes 40% less energy, is biodegradable and it has reduced greenhouse gas emissions by 20% compared with petroleum-based propanediol.

As oil supplies for petrochemicals becomes more expensive or restricted due to targets for net zero emissions, the use of biotechnology to produce chemicals will grow. Bioethanol is a low-carbon, renewable source of energy used in transport fuels that has lower greenhouse gas emissions and improved air quality. Bioethanol (or biodiesel) is made from vegetable oils. Biodegradable polymers such as polylactic acid (PLA) for packaging are made from plant starch. Historically, ethanol derived from fermentation of sugars provided fuel for early cars before fossil fuels were commonplace including the Ford Model T could be run on ethanol in the 1910s. Currently most bioethanol is used in mixtures with petrol. In 2007 British Sugar operated the first bioethanol factory in the UK at Norfolk, making about 70 million litres of renewable bioethanol from locally grown wheat and sugar beet.

Synthetic natural gas can be produced from anaerobic digestion of non-woody biomass by micro-organisms in the absence of oxygen to produce methane gas. The biogas can be combusted to generate heat and power or further processed into a liquid fuel. The digestate waste forms a nutrient-rich material used to improve soil. Research is also focusing on the potential to produce meat and other food proteins in bioreactors. Industrial biotechnology will play an increasing role in the production of many chemical products in the future, with the global industrial biotechnology market expected to grow to over $800 billion by 2030. It will serve a diverse range of new and existing industries, including food and beverage, pharmaceuticals, renewable energy, fuels, numerous interesting materials and speciality chemicals.

Z: Zero greenhouse gas emissions

Finding the right balance

Achieving 'net zero' to reach an overall balance between the greenhouse gas emissions produced and those taken out of the atmosphere is seen as an important part of reducing the impact of long-term climate change. Certain atmospheric gases let sunlight pass in but keep radiated heat from escaping, this is analogous to the glass panes of a greenhouse; hence they were named *greenhouse* gases. The 'heat-trapping' nature of carbon dioxide was first demonstrated in the mid-19th century. New chemical technologies are providing the means of reducing current emissions, actively removing greenhouse gases from the atmosphere and supporting adaptations to the new climate conditions.

Great technological, economic and social progress came about as a result of industrialization, which first began in Britain in the 18th century powered by readily available fossil fuels such as coal, oil and gas. There were untold benefits that came – growing wealth, the ability to feed and house expanding populations living in cities, new advances in science and engineering, enabling new materials for the manufacture of consumer goods, generating energy for powering machines and electrical goods, and making fast transportation by road, air and sea widespread. The great benefits from the combustion of the fossil fuels also gave rise to the emission of greenhouse gases such as carbon dioxide, driving up global temperatures with potential consequences for the environment. Companies and countries are now focusing on the transition to a 'net zero' economy and adapting to the changing climate.

Climate in history

In the past 800,000 years there have been multiple cycles of glacial advance and retreat, with the abrupt end of the last ice age about 11,700 years ago marking the beginning of the modern climate era and of human civilization. In the past, climate change can be attributed to very small variations in earth's orbital position altering the amount of solar energy it receives. There have been periods of warming before, such as in the 'medieval warm period' from 800 to 1250, and significant declines in temperature during the 'Little Ice Age' from 1410 to 1720. A decrease in solar activity coupled with an increase in volcanic activity that emitted particles, temporarily blocking sunlight, is thought to have helped trigger the Little Ice Age – the coldest period since the last full ice age. In 1991 a large eruption in the Philippines reduced average global temperatures by 0.5 °C for more than a year as dust particles deflected sunlight. Regular large-scale weather patterns such as El Niño in the eastern Pacific also have a significant global impact on climate.

Before industrialization, carbon dioxide mainly came from natural sources including respiration, plant decay and volcanoes. As plants grow, they take carbon out of the atmosphere, and when they decay it is released again. Millions of years ago some carbon from decaying organic matter became transformed into layers of coal, oil and natural gas. By burning these fossil fuels, millions of years of carbon uptake by plants has been returned to the atmosphere in less than 300 years. Historical evidence for variations in levels of greenhouse gases and temperature comes from examining samples of ice cores drawn from Greenland and Antarctica, from tree ring growth, ocean sediments, coral reefs and layers of sedimentary rocks. The gases can remain in the atmosphere for long periods and have a worldwide impact, no matter where they were first emitted. There is evidence that the amount of carbon dioxide in the atmosphere has increased by more than a third since the Industrial Revolution from a concentration of 280 parts per million (ppm) in 1750 to a global average in 2020 of 410 ppm. Annual human-generated carbon dioxide emissions from combined fossil fuel consumption and land-use activities were just below 40 billion tonnes in 2020. Consequently, it was reported in 2022 that the earth was about 1.2 °C warmer than the late 19th century average when modern recordkeeping began. In the past 50 years earth-orbiting satellites have enabled scientists to collect information about the planet and its climate on a global scale. Temperatures have not risen everywhere or in every year by the same amount as warming differs substantially on certain land masses and over oceans.

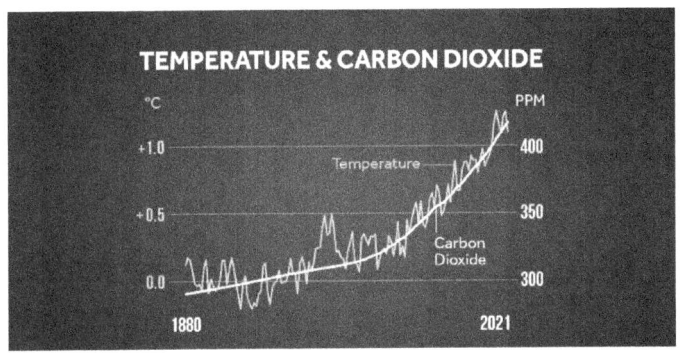

Average global temperature and carbon dioxide levels (compared to 1880-1900)

In 2015 governments around the world agreed a legally binding international treaty in Paris on combating climate change and adapting to its effects. This aims to limit the global temperature increase to below 2 °C compared with pre-industrial levels by reducing anthropogenic greenhouse gas emissions while helping countries to adapt to the impacts of climate change. Many technological changes are required to reduce emissions of greenhouse gases and achieve net zero emissions by the middle of the 21st century. Among these are the phasing out of fossil fuel consumption by transitioning to renewable forms of energy such as solar and wind, achieving greater energy efficiency and adopting new means of energy storage and transport. Changes in chemical technologies are required to 'decarbonize' industrial manufacturing processes, particularly in those with the highest emissions such as steel and cement production, plastics or fertilizers. It is still possible to produce some emissions if they are offset by other actions that reduce greenhouse gases through natural carbon 'sinks', such as the oceans, making changes in land use by planting forests and crops or using carbon emissions to make new products. It will also be possible to create artificial underground sinks to store carbon.

Globally, energy demand is expected to increase by 35% before 2030, although greenhouse gas emissions from energy use may have almost peaked due to the ongoing transition to renewable energy sources. Although most of the cumulative carbon emissions since the Industrial Revolution have occurred in Western countries, now other areas, such as Asia Pacific, that have grown rapidly in population and income account for around 70% of the current emissions.

In the UK the total greenhouse gas emissions were 48% lower by 2021 than they were in 1990 at 427 million tonnes. UK emissions have fallen due to the rapid uptake of renewable energy, the phasing out of coal-fired power generation and a reduction in heavy industrial activity. Although Britain's total emissions make no more than a very small contribution to global emissions (less than 1%), it was the first major economy in the world to pass laws to end its contribution to global warming by 2050. These commit it to investments in innovative climate technologies and infrastructure to transition away from fossil fuels.

Pioneers of the 'greenhouse effect'

Several early scientists had investigated this effect, such as Joseph Fourier, who in 1824 had calculated that an earth-sized planet, at our distance from the sun, ought to be much colder unless the atmosphere was acting like an 'insulating blanket'. An American amateur scientist, Eunice Foote, postulated that it was small amounts of carbon dioxide and water vapour in the atmosphere that were responsible. The physical basis of this effect was first demonstrated in 1859 by the Irish physicist John Tyndall (1820–1893), who established that carbon dioxide and water vapour in the atmosphere were among the gases that absorbed heat radiated from the sun, impacting the weather and potentially changes in climate. He wrote: '... *the atmosphere admits of the entrance of solar heat; but checks its exit, and the result is a tendency to accumulate heat at the surface of the planet.'* Tyndall demonstrated the absorption and radiation by certain gases of long-wave infrared radiation. Of course, without the natural warming from this effect, when the atmosphere traps heat radiating from the surface back towards space, the earth would not be habitable for much life.

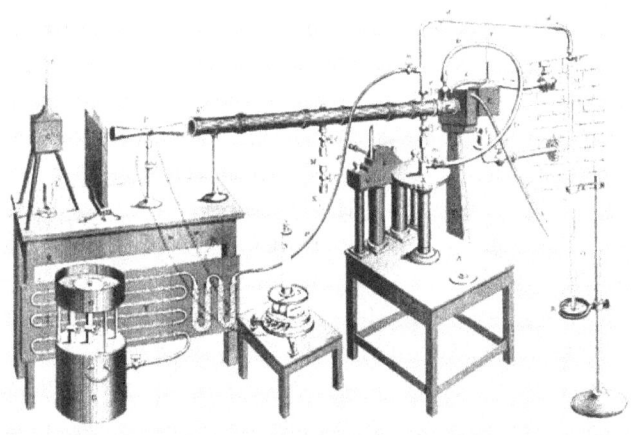

Apparatus for measuring the heat absorption of gases

Svante Arrhenius (1859–1927), who was born near Uppsala, in Sweden, studied electrochemistry at Uppsala university and was awarded the Nobel Prize for Chemistry for his theory of electrolytic dissociation. In 1891 the Swedish chemist also founded the Stockholm Physical Society for a group of scientists whose interests included geology, meteorology and astronomy. This group stimulated his interest in the physics of the earth, sea and atmosphere. In 1896 he calculated that changes in atmospheric carbon dioxide levels could substantially alter the surface temperature. Arrhenius's approach was to make sense of existing observations and experiments by other scientists, while trying to understand the cause of ice ages. He presented a paper to the society entitled *On the influence of carbonic acid (carbon dioxide) in the air upon the temperature of the ground*. Here he described an energy budget model that considered the radiative effects of carbon dioxide and water vapour on the surface temperature of the earth, and variations in atmospheric carbon dioxide concentrations. Arrhenius suggested that variations in carbon dioxide levels in the atmosphere could greatly influence the heat budget of the earth. He performed a series of calculations on the temperature effects of increasing and decreasing amounts of carbon dioxide in the earth's atmosphere. He showed that the temperature of the Arctic regions would rise about 8–9 °C if the gas increased 2.5 to 3 times its present value. The low temperatures experienced in the Ice Age occurred at about half of the present value, which would lower the temperature by 4–5 °C. As Arrhenius predicted, both carbon dioxide levels and temperatures did increase from 1900; it was less than his forecast, but he had identified the trend. In 1938 Guy Callendar, a Canadian engineer with an interest in climate, first presented evidence that the increase in global temperatures could be the direct impact of the increase in carbon dioxide emissions from industrial activities.

Present climate research attributes the global warming trend observed since the mid-19th century to a range of human activities that have impacted the greenhouse effect. Such activities include burning of fossil fuels and others such as population growth, deforestation, intensive agricultural practices, building large urban areas and the expansion of industrial activities.

While carbon dioxide represents 80% of all greenhouse gases emitted, others are emitted in smaller quantities, some of which trap heat far more effectively. Methane is a more potent greenhouse gas than carbon dioxide by a factor of 30 times over a long period. It is produced through drilling for oil and gas but is also emitted in agriculture from ruminant digestion and manures associated with domestic livestock, growing rice or from the decomposition of waste. Methane makes up about 10% of emissions. Nitrous oxide is another powerful greenhouse gas produced in agriculture from soil cultivation practices, especially the use of commercial and organic fertilizers, burning fossil fuels, decomposing biomass or from nitric acid production, making up about 5% of total emissions. Water vapour too is an abundant greenhouse gas, but it usefully operates as a 'feedback control' because it increases as the atmosphere warms, leading to clouds and precipitation, which in turn reduce the solar radiation. Natural sinks such as oceans, forests, plants and chemical processes that absorb or store carbon dioxide from the atmosphere remove some 10 gigatonnes of the gas per year, a third being absorbed by the oceans.

Consequences of global warming

In some places warmer temperatures could be beneficial in improving crop growth and widening the range of crops that can be grown, but elsewhere hotter conditions may result in more arid or seasonally flooded areas. While higher atmospheric carbon dioxide increases crop yields, clearly weather extremes, such as droughts, floods and high temperatures, can cause physical crop losses, reducing farm incomes and threatening food security. Depending on the crop types and ecosystem, climate change may cause new patterns of pests, diseases and weeds to emerge. As the oceans are absorbing much of the energy accumulated, they are becoming warmer in the surface layer. The carbon dioxide that they absorb causes the acidity of sea waters to increase, promoting algae and seagrasses, sometimes harmful to oysters, clams and coral. Warming the oceans also decreases their ability to absorb more and could lead to degassing of some already dissolved carbon dioxide.

With warmer air and water, sea levels will rise as the water expands and more ice melts from ice sheets in Greenland and Antarctica. The global sea level rose by about 200 mm in the 20th century and it is currently rising by about 25 mm per decade, which leaves many low-lying parts of the world vulnerable to flooding, particularly near riverine and coastal settlements. The increased ocean temperatures can begin to destabilize glaciers that flow into the oceans. Glaciers have been seen to retreat from mountainous regions too, so there is less snow cover and earlier snow melting. Arctic sea ice has declined by about 13% per decade since 1980, which may result in ice-free summer periods. Retreating sea ice means more heat is absorbed as the darker sea surface reflects only 7% of sunlight while snow-covered sea ice can reflect about 85% of the sunlight back into space. Clearly the less sunlight the earth's surface reflects, the more heat is absorbed by the oceans, which heat up, melting even more ice. Less sea ice will unlock northern shipping routes and access more natural resource reserves, but the Arctic ecosystems remain fragile. When Arctic land thaws, once frozen ground (known as 'permafrost') releases more carbon dioxide and methane gas. The rise of temperatures in the Arctic might change atmospheric circulations, including the path and speed of the jet stream, leading to periods of more unusual weather conditions. Of course, computer modelling of the atmosphere does not represent the variability well and all future climate predictions remain subject to large uncertainties.

Adapting for net zero

The global focus is on following a net zero pathway with large-scale reductions in the use of fossil fuels by making investments to build a resilient economy that employs renewable energy such as wind, solar and tidal power and extensive electrification, particularly for transport and heating, plus the use of low-carbon power generation methods such as bioenergy powered by waste or nuclear energy. Nuclear power plants produce no greenhouse gas emissions during operation and provide a baseload to complement variable solar-powered generators. Further, the development of a hydrogen-based economy to meet the demands of energy-dense chemical processes and fuel long-distance heavy road and sea transport will also be required. There is an urgent need to improve resource and energy efficiency to reduce total demand.

By far the most important response to climate change is adapting (for greater resilience) the built environment and its infrastructure. Extreme weather events, such as storms and floods, cause disruption or sometimes the complete loss of essential services such as water, public sanitation and energy supplies, or create problems in transportation and communication networks. As well as being costly to recover, the loss of these services can have significant impacts on people's safety, healthcare and economic wellbeing. In the UK, for example, the increasing frequency and severity of flooding represents the most significant risk to infrastructure and homes. Many existing buildings and homes, coastal and inland waterways and roads will be modified or protected to cope with the changing conditions. There are many chemical technologies that can support making products for the more resilient designs that are required.

Contribution of chemical technology

Currently, the chemical sector is the largest industrial consumer of oil and gas, both as a chemical feedstock and as energy. The chemical sector's high energy consumption is propelled by demand for the vast array of chemical products, which is expected to double globally by 2050. In addition, the traditional chemical technologies for making cement, iron and steel, and fertilizer produce significant carbon dioxide emissions. Chemical and related industries across the world are working hard to reduce material yield losses. For example, currently 25% of all steel and 50% of all aluminium never makes it into the final product but must be re-melted, which wastes energy and produces unwanted emissions. China, where much of the global chemical, steel and building materials are produced, is still a massive user of carbon-intensive coal.

Chemical technologies have an important role in reducing emissions by creating new efficient chemical processes, energy-saving insulating materials, processes for recycling and components for renewable energy technologies. The latter includes developing battery and energy storage or producing electrolytic hydrogen and storage systems. Many insulation materials made for homes and other uses save over 2 tonnes of greenhouse gases for every 1 tonne they directly emitted in production. Heating homes and buildings is responsible for almost 20% of the UK's carbon emissions. Chemical producers face a particular challenge in reducing emissions, due to the long lifespan of existing equipment such as oil crackers or the need to generate high temperatures for many processes, which are impractical to achieve with electricity. There is also the problem that retrofitting climate-abatement measures in chemical plants will often raise costs as they reduce throughput and require massive up-front investment in recycling equipment or carbon dioxide capture. The production of primary chemicals such as ethylene, propylene, benzene, toluene, xylene, methanol and ammonia accounts for two-thirds of energy consumption in the chemical and petrochemical sector. Demand for essential plastics used in packaging, construction and automotive applications is also driving demand for these primary chemicals, as they are the key precursors to most of these plastics. Many important chemicals cannot rely on decarbonization but will require alternative carbon raw materials. Most countries have banned single-use plastic items such as plastic bags, cutlery and straws and have regulations to ensure that plastics must contain more recycled material. Recycling counterbalances a proportion of global demand for new plastics and reduces waste. Plastic wastes can now be reprocessed using special innovative chemical plants into other useful materials. An aviation fuels technology company (Lanzajet) has developed an alcohol-to-jet technology for the commercial production of both sustainable aviation fuel and renewable diesel from waste ethanol, produced from municipal solid waste, agricultural residues, industrial off-gases and biomass fermenters. This could lead to a real reduction in emissions in the important air-travel sector. Chemical production can be decarbonized by replacing fossil fuel raw materials with chemicals made from recycled carbon dioxide, water and electricity. One US research group (Renew CO_2) has developed a proprietary catalyst that enables a low-temperature process for converting carbon dioxide into ethylene glycol, a monomer for polyester fibre.

Synthetic nitrogen fertilizers used in half the world's food production are made from ammonia, which is a very large-volume and energy-hungry chemical, so modern farming is focusing many efforts to increase nutrient efficiency and employ good carbon management (see N). Hydrogen is going to have an important role in the future as, unlike methane or other fossil fuels, it contains no carbon and so when it is burned it produces only water vapour. It will be important in some industrial processes and energy-dense applications such as long-distance heavy goods vehicles and ships, or for storing the gas for rapidly producing electricity and heating in peak demand periods. Hydrogen can be used as a precursor for other 'energy carriers', as a feedstock for ammonia manufacture or other synthetic hydrocarbons or to directly power fuel cells in cars and ships. Toyota produced the first mass-produced mid-size hydrogen fuel cell vehicle (FCV), the *Mirai* car, in 2014. While new chemical technologies are being pursued to produce hydrogen, they remain at the early stages of development.

Toyota Mirai hydrogen fuel cell car (2016)

Unfortunately, hydrogen although widely used as an industrial chemical, is currently either made through the gasification of coal or through steam methane reformation using natural gas as the feedstock. The hydrogen is considered 'green' when made from electrolysers powered by renewable electricity. This green hydrogen can cost two to three times more than that derived from methane. The other problems in using hydrogen concern safely storing and transporting it, as it is a highly flammable gas, which takes up a lot of space (due to its low density) and can cause steel pipes and welds to become brittle or fail. The bulk transport of hydrogen will require dedicated pipelines and pressurizing or cooling it to a liquid. A British scheme (HyNet) is underway to provide the infrastructure to produce, transport and store low carbon hydrogen across the North West England.

Carbon removal

It is unlikely that emission reduction or elimination alone will achieve net zero carbon unless this is combined with methods of carbon removal or 'capture' from the environment. Net zero will require emissions to be balanced by schemes to offset an equivalent amount of greenhouse gases from the atmosphere, such as planting forests and woodland trees or increasing soil's ability to capture carbon. A new woodland takes over 20 years to absorb significant carbon dioxide, while farmland crops absorb twice the annual quantity of carbon dioxide as a mature woodland. Improvements in management of the water-logged soils used in rice paddy cultivation can significantly reduce methane emissions. In the UK, a basalt rock dust is being spread on fields to capture carbon from air when the dust weathers. There are emerging new technologies such as carbon capture, utilization and storage (CCUS) that provide a way of reducing carbon emissions by capturing the carbon dioxide produced by power generation or industrial activities, such as making steel or cement. The gas is then transported to be stored deep underground in geological formations such as saline aquifers or depleted oil and gas reservoirs. Globally some 2 gigatonnes of carbon may need to be captured each year by 2050. As already mentioned, instead of storing carbon it could be reused in new industrial processes by converting it into, for example, plastics, resins, concrete or biofuels. A new chemical technology (from Planetary Technologies) is under investigation to remove carbon dioxide from air, generate hydrogen and reverse ocean acidification in an integrated process.

Difficult challenges ahead

Although there have been concerted international efforts over the past 20 years to increase the amount of electricity generated by wind, solar and other renewable sources, it should be recognized how dependent the world is on fossil fuel feedstocks for power, transport and manufacturing essential consumer goods. Absolute global emissions of greenhouse gases have continued to rise by about 40% (from the1990 baseline) as global growth continues. The UN estimates that to limit increases in global temperature to less than 2 °C, only about another 1,000 billion tonnes, or about half of what has been emitted so far, can still be emitted. This remaining emission budget would be used in less than 30 years.

Energy production from fossil fuels is still being used to produce 60% of electricity globally. There are serious problems relying on renewable energy when the sun does not shine or there is no wind. While it is possible to store electricity there can be periods with long wind lulls, with the sun low in the sky or night-time necessitating huge quantities of expensive storage. In the medium term, decarbonization of the global economy is going to result in new environmental costs and taxes to support funding capital expenditure. The UK government's independent climate adviser, the Climate Change Committee, estimated that to reach net zero emissions by 2050 the UK will require investments costing over £1,000 billion, some of which may be offset by energy savings. In many countries, voters, tax payers and consumers are being asked to pay more for energy and goods to give benefits to the environment they may not live to enjoy, and this is sometimes negated by other countries failing to take similar action on reducing emissions. The drive to net zero is leading to continuing deindustrialization, with the closure of traditional manufacturing industries such as blast furnaces, oil refineries, aluminium smelters and fertilizer plants. The UK's oil capital Aberdeen is already moving away from an industry that has created half a million jobs and long brought wealth to north-east Scotland. UK oil production peaked in the 1990s from over 2.5 million barrels per day, falling over 25 years to less than 700,000 barrels daily, with a significant impact on employment so new roles are required in renewable energy or alternative industries. Ironically, fuels are now being imported and manufactured goods once produced in the UK are sourced from China and other countries resulting in no actual carbon savings.

Steven Koonin, physicist and former advisor in the US Department of Energy reminds us: *Climate is a 30-year average; weather is what happens every day or every year. They are not the same thing. People confuse climate change with a changing climate. Climate change has come to mean changes due to human influences.* Many politicians, journalists and academics have distorted the actual situation to make unsubstantiated predictions and promote unrealistic actions.

Fortunately, some new sustainable technologies have become cheaper than conventional technologies and represent good investments. Chemistry and chemical technology are playing leading roles in tackling the major challenges of climate change, particularly in producing high-performance materials that do not require fossil fuels, storing energy in banks of batteries and driving the development of a range of new processes for alternative fuel sources such as hydrogen. New chemical processes are also emerging for recycling materials, using chemical waste products, creating new energy-saving materials and facilitating carbon capture, usage and storage. Many important chemicals cannot rely on decarbonization but will require alternative carbon raw materials to be developed. In the future, these areas will grow in importance and chemistry and chemical technology will contribute massively to their success.

5. Epilogue

Being part of the chemical technology future

This book has featured many of the important chemical and some biochemical technologies created over the past few centuries. What chemical technologies will feature in your lifetime? Bearing in mind the chemical industry has undergone huge changes in the past 50 years and it is now a totally global, diverse and complex enterprise – it is difficult to forecast.

Sustainable development involves balancing the protection of the natural environment with meeting human economic needs over the long term. In the 21st century the chemical industry faces many challenges as it helps to maintain and improve standards of living in a sustainable way. Chemical technologies are now being directed at reducing the costs of products, using new energy sources, adapting to the impacts of climate change, growing more food, improving population health while minimizing pandemic diseases and dealing with multiple threats to global security. The civil infrastructure faces threats from ageing, extreme weather events and terrorist attacks. The chemical technologies of the future will need to preserve ecological balances by providing solutions that are feasible technologically but also economical when compared with using the fossil fuel technologies of the past.

As the global population continues to grow towards 9 billion by 2040, ensuring adequate energy, food and fresh water with continual advancement are key goals. Previously the oil and petrochemical industries provided excellent energy sources, an array of raw materials, polymers and many sophisticated downstream chemicals at a relatively low cost. The key to the future is finding ways to reduce dependence on non-renewable resources as supplies dwindle, lowering greenhouse gas emissions and improving energy efficiency. Reducing the accelerating appetite for rare metals and minerals required in electronic devices and batteries will be essential too. As they did in the past, the world's chemists, chemical engineers and other technologists are finding new chemical solutions to these multiple challenges. In the late 19th and early 20th centuries, the chemical industry was based largely on coal, ores containing metals, minerals and natural biomass. A return to older technologies to produce chemicals in an environmentally acceptable way may be possible – for example, ethene, ethanol and a range of polymers can already be produced from biomass.

Biodegradable packaging made from artichokes

Historically, as large energy users, many chemical technologies may not have been associated with positive environmental outcomes, but now it is clear that net zero and sustainable economies cannot be achieved without the contribution of the chemical industry such as energy-saving building materials. To reduce dependence on non-renewable energy resources requires the redesign of chemical plants to run at lower temperatures, using catalysts or finding radically alternative chemical routes. In recent times, after some deindustrialization and the switch to more renewal sources, the consumption of energy and emissions per unit of production has fallen in European countries by about 55%, since 1994. Chemical technologists are supporting the development of carbon-free, renewable energy solutions, involving biofuels, electric batteries, fuel cells and nuclear power generation.

While the sun is radiating to earth more energy each hour than the whole population consumes in a year, capturing that power, converting it into useful forms and especially storing it poses significant engineering challenges. Anticipating some continued use of fossil fuels, chemical engineers are exploring technological methods of capturing the carbon dioxide produced from fuel burning, using it in the production of new materials or sequestering it underground.

Water is essential to life on earth but is in seriously short supply in many regions of the world. The Food and Agriculture Organization predicts that by 2025 nearly 2 billion people will be living in areas of absolute water scarcity, and two-thirds of the world's population will live in water-stressed areas. Clean water for drinking and cooking is a limited resource that is coming under increasing pressure through population growth, industrialization and agricultural intensification. As in the past, chemical technologies have a huge role to play in treating and recycling wastewater or making industrial and communal water use more efficient. New large-scale technologies for desalinating seawater may be helpful, but small-scale technologies for local water purification will also be needed to meet personal hygiene needs. In addition, advanced treatment processes now make it possible to recover valuable materials from wastewater, including metals, nitrates, phosphates and biogases.

In the past, synthetic fertilizers have allowed global food production to broadly keep up with population growth, but the increasingly limited availability of suitable land, shortages of farm labour and again water, coupled with the impact of climate change, threatens to disrupt this. As agriculture accounts for 30% of greenhouse gas emissions, new farming methods and chemical technologies are needed to conserve water, optimize crops yields and reduce fertilizer and pesticide use. With widespread use of fertilizers and industrial combustion, humans have doubled the rate at which nitrogen is removed from the air relative to preindustrial times, contributing to smog and acid rain, polluting drinking water and impacting global warming. Consequentially, there is a need for counter measures against these effects while maintaining the ability of agriculture to produce increasing food supplies.

Over the 20th century, chemical and biological technologies ensured many of the health problems and contagious diseases of the past were controlled and even eliminated. Other old diseases such as malaria remain deadly, while newer health problems such as dementia and obesity are yet to be fully addressed by medicine. New approaches promise 'personalized' medicine. Doctors have long recognized that individuals differ in their susceptibility to disease and their response to treatments, but traditionally medical technologies have offered 'one size fits all' remedies. The cataloguing of the human genome and deeper understanding of the body's complement of proteins and their biochemical interactions offer the prospect of identifying the specific factors that determine sickness and wellness in any individual. Certain deadly bacteria, for instance, have repeatedly evolved new properties, conferring resistance against even the most powerful antibiotics. As 21st-century pandemics have shown, new harmful viruses can arise, which in the highly connected modern world spread more rapidly than traditional disease-prevention measures.

Consumers have become increasingly focused on the environmental impact of the products they buy. There have been bans on single-use plastics, but it was the great innovations in plastic materials in the past, from Bakelite or polyethylene to PTFE, that transformed modern living. There are opportunities for the development of new alternative materials, such as biopolymers. Potential innovations in materials for 3D printing, nanomaterials or graphene technology present great hopes for producing new generations of raw materials for the industries they serve. Producers are partnering with suppliers and customers on initiatives for chemical recycling and creating 'circular' supply chains. It is possible to thermochemically recycle plastic waste and produce hydrocarbon feedstocks for reuse in downstream processes. There is a recycling project at BASF (*ChemCycling*) to manufacture high-performance products from recycled plastic waste. Also on the increase are plants supplying power utilizing energy-from-waste (EfW) units. In Sweden only one percent of household waste goes to landfill as millions of tonnes are safely incinerated to produce power.

The large scale, flammable nature or special features of chemical assets mean that managing safety and risk in chemical engineering is often very different from managing risk in other industries. Designing and operating high-hazard facilities, where accidents are rare but can have devastating impacts, demands an exacting approach to safety and loss prevention. This requires skilled professionals, effective managers and a well-trained workforce coupled with investments in protective control and safety systems backed by regulations. The chemicals sector will continue to have a high dependency on measurement, process control and predictive analytics for automatic processing in the 'smart' chemical factories of the future.

All these examples merely scratch the surface of the exciting challenges and complexity of the tasks that chemical technologists will face in the future. Science and innovation are the bedrocks of British industry and its economy. That excellence has been witnessed throughout the centuries, in early textile manufacturing or chemical innovation, astronomy, aerospace, motor racing, creation of fuel cells and the World Wide Web, health sciences, the discovery of penicillin or rolling out the world's first approved COVID vaccine. There is no escaping the fact that countries such as the UK are in a race with other science superpowers. A sense of urgency is needed to capitalize on that expertise as huge investment decisions are being made worldwide. The UK retains a reputation as a global leader in science and for the quality of its researchers, but often its start-up companies move abroad to access funding. Some governments are giving their companies significant support for investments in clean energy, electronics, biotechnology and healthcare. This promotes research and development in state-of-the-art chemical technologies, including batteries, green hydrogen, biotechnology, carbon capture and storage.

Those hoping to build a future workforce in chemical technology must co-operate with companies, educators, scientists and engineers to encourage and promote improved science, technology, engineering and mathematics (STEM) education in schools. They should ensure positive messages reach students and the public at large, conveying not just the facts of science and engineering but also an appreciation of the ways that scientists and engineers work, how they can benefit society and not forgetting mention of those many pioneers of the past. Chemical and engineering research is employing new emerging tools such as AI, molecular modelling, quantum chemistry and synthetic biology. Industrial chemists and chemical engineers can bring a unique perspective as they are trained to think broadly, work across many disciplines and understand whole processes, unit operations and systems.

Having considered the question at the beginning of this book about what use is chemistry, you might hopefully now conclude it is a great time to think about following a career in chemistry, the biosciences or chemical engineering. Gaining an understanding of some of the important chemical technologies and pioneers of the past as described in this book will hopefully help you find new perspectives and motivation to start pursuing this century's great challenges. Chemical technologists of the future will make the world not only a more technologically advanced but also a more sustainable, safer, healthy and better place to live. *Perhaps you are going to become one of those chemical technologists?*

Acknowledgements

I was inspired to write this book partly by my early experiences of technology often gained by reading about the pioneers of the past. From an early age I spent hours formulating products to test on my sister, Alex (not always appreciated); conducting experiments from a large chemistry set; setting off rockets in the garden and constructing crystal and transistor radios. It was at the time of the 'Space Race' featuring the exciting Apollo missions to the moon. My father, an aero-engineer gave me an early awareness of industrial engineering. Once at high school my interest in chemistry and physics increased. I learnt about the great British scientists and engineers of the past such as Davy, Dalton, Faraday, Clerk-Maxwell, Kelvin but also enjoyed reading about pioneering business and technology entrepreneurs such as Edison, Nobel, Carnegie, the Wrights, Marconi and Ford.

I often found that an initial spark came from seeing something about the history of technologies past or its pioneers which then ignited a passion to explore. Such sparks and a like for making things led me into a career in the chemical industry, initially as a chartered chemical engineer, then later in manufacturing and business management where I was grateful to be continually inspired by many colleagues, customers, chemists, chemical engineers, technologists and technicians who made working in industry varied, challenging, fascinating and often enjoyable. Thank you one and all.

To produce this non-profit making book, I am very thankful to Robin Bell for generously sponsoring the project through his charity: *The Robin and Eirwen Bell Trust* which encourages young people to engage in the arts, heritage, culture, health and science.

I am delighted that Steve Bagshaw CBE, a former business and manufacturing colleague has generously provided the foreword to the book. He was chief executive officer for six years of Fujifilm Diosynth Biotechnologies, a globally leading manufacturing and development organisation. Having almost 40 years distinguished experience across the biotechnology and fine chemicals business and manufacturing sectors he is well placed to see the importance of the chemical and biological sciences.

Many thanks to several experienced technologists including Andy Collins a former senior industrial chemist, and secondary teacher who kindly made useful comments on my manuscript. Advice from Robin Bell, himself an established writer, author and historian was much appreciated. NGP prepared the book for publication. Neil Thompson at Danscot designed the cover.

Finally, I am, as always, very grateful for continued encouragement and support from my talented wife Karen, herself an experienced industrial research and development chemist, a former technical and operations manager in the chemical industry and now an accomplished secondary school chemistry teacher. I dedicate this book to her.

Abbreviations

ABS	acrylonitrile butadiene styrene
AFC	alkaline fuel cell
AI	artificial intelligence
ANFO	ammonium nitrate fuel oil
BASF	Badische Aniline und Soda Fabrik
BOS	basic oxygen steelmaking
BOV	brown oil of vitriol
BP	British Petroleum
CCUS	carbon capture, utilization and storage
CFC	chlorofluorocarbon
DBP	disinfection by-products
DMFC	direct methanol fuel cells
DNA	deoxyribonucleic acid
DRI	direct reduced iron
EAF	electric arc furnace

EfW	energy-from-waste
EPA	Environmental Protection Agency
EV	electric vehicle
FCC	fluidized-bed catalytic cracking
FCV	fuel cell vehicle
GMO	genetically modified crops or organisms
HDPE	high-density polyethylene
HFA	hydrofluoroalkane
HID	high-intensity discharge
HPV	human papillomavirus
IC	integrated circuit
ICI	Imperial Chemical Industries
IPV	inactivated poliovirus vaccine
ISSP	Institute of Solid State Physics
LCD	liquid crystal display
LDPE	low-density polyethylene
LIB	lithium-ion battery
LFP	lithium iron phosphate battery

MIT	Massachusetts Institute of Technology
MRI	magnetic resonance imaging
MRSA	methicillin-resistant *Staphylococcus aureus*
NMR	nuclear magnetic resonance
NSAID	non-steroidal anti-inflammatory agent
OPV	oral poliovirus vaccine
PAFC	phosphoric acid fuel cells
PCCO	praseodymium cerium copper oxide
PCR	polymerase chain reaction
PEMFC	proton exchange membrane fuel cells
PET	polyethylene terephthalate
PHA	polyhydroxyalkanoates
PLA	polylactic acid
POP	persistent organic pollutants
PPI	proton-pump inhibitor
PTFE	polytetrafluoroethylene
PtL	Power-to-Liquids
PV	photovoltaic

PVC	polyvinyl chloride
RNA	ribonucleic acid
RO	reverse osmosis
ROV	rectified oil of vitriol
SAF	sustainable aviation fuel
SARS	severe acute respiratory syndrome
SBR	styrene-butadiene rubber
SMR	steam methane reforming
STEM	science, technology, engineering and mathematics
TCC	Thermafor catalytic cracking
TNP	trinitrophenol
TNT	trinitrotoluene
UAC	United Alkali Company
UK	United Kingdom
UPS	uninterruptible power supplies
US	United States
UV	ultraviolet
VLP	virus-like particles

Selected Bibliography

The History of Chemistry *John Hudson;* The Macmillan Press 1992

Shaping the Industrial Century *Alfred Chandler;* Harvard University Press 2005

The Chemical Industry *Alan Heaton;* Chapman and Hall 1994

British Chemical Industry *Gilbert Morgan and David Pratt;* Edward Arnold 1938

A History of the British Chemical Industry *Stephen Miall;* Benn 1931

A History of the International Chemical Industry *Fred Aftalion;* Chemical Heritage Foundation 2001

A History of the Modern British Chemical Industry *Hardy and Davidson Pratt;* Pergamon 1966

The Awakening Giant, *Andrew Pettigrew;* Basil Blackwell 1985

Concise Encyclopaedia of Chemical Technology Kirk-Othmer; Wiley-Interscience 1999

The Progress of Science in the Conquest of Disease: I C (Pharmaceuticals) Ltd 1954

History of the International Dyestuffs Industry *Morris and Travis;* American Dyestuff Reporter 1992

The Dyes: Scotland's Dyestuff Pioneers and a Century of Manufacturing *John Blackie;* DP 2023

Mauve *Simon Garfield;* Faber & Faber 2013

Development of the UK Chemical Industry *David Culpin;* CIA 2000

Dyes and Textiles in Britain *James Morton;* British Association 1930

James Kier: A Renaissance Man of the Industrial Revolution *Kristen;* Schranz 2017

A History of Widnes *Revd. G. Diggle;* Corporation of Widnes 1961

Discovery of Polypropylene and High-Density Polyethylene; American Chemical Society 1999

Civil Engineering Materials *Peter Claisse;* Butterworth-Heinemann 2015

Making the Modern World: Materials and Demineralization *Vaclav Smil;* John Wiley and Sons 2014

Makers of the Microchip: Fairchild Semiconductor *C. Lecuyer and D. Brook;* MIT Press 2010

Understanding Semiconductors: A Technical Guide *Corey Richard;* Apress 2022

Carbon Capture and Utilization in the Steel Industry *Kevin de Ras;* Chemical Engineering 2019

Lithium-Ion Batteries: Basics and Applications *R. Korthauer ed;* Springer 2018

The role of the medicinal chemist in drug discovery *J.G. Lombardino;* Nature Reviews 2016.

The Unbearable Cost of Drug Development; Genetic Engineering & Biotechnology News 2023

What is medicinal chemistry? *S. Holbrook;* RSC September 2017

Enriching the Earth: Transformation of World Food Production *Vaclav Smil;* MTI 2001

The Wheat Problem: Presidential Address British Association *W. Crookes;* John Murray 1899

Modern Production Technologies: Ammonia et al: a review *Max Appl;* CRU Publishing 1997

Billingham: The First Ten Years *V. E. Parke;* ICI Billingham 1957

Ammonia Technology Development from Haber-Bosch *J. G. Reuvers;* Proceedings IFS 2014

The Alchemy of Air *T. Hager;* Harmony Books 2008

A New Process for Large-Capacity Ammonia Plants *Larsen J*; Nitrogen & Methanol 2001

Nobel Prizes and Laureates: Nobel Prize in Chemistry 1931; The Nobel Foundation 1931

Introduction of the Pill and its Impact. *L.Tyrer;* Contraception 1999

International Society for Plant Pathology; Nature Ecology & Evolution March 2019

Concerns Over Use of Glyphosate-based Herbicides *John Myers;* Environ. Health. Feb. 2016

The Evolution of Clean: a visual journey *Fortuna Spitz;* Soap and Detergent Association 2006

Short History of Soap *John Hunt;* Pharmaceutical-Journal RPS 1999

The History of the Manufacture of Soap *Gibbs F.W.;* Annals of Science 1939

Biography of William Lever (Lord Leverhulme); The Chemical Age 1919

Poucher's Perfumes, Cosmetics and Soaps *Michael Willcox*; Chapman & Hall 1993

Soap through the Ages *R. Lucock Wilson* R; Unilever Limited 1957

The Science and Technology of Rubber 4th Edition *Ed James E. Mark* 2013

Rubber Survey Ullmann's Encyclopaedia of Industrial Chemistry *H. Greve;* Wiley-VCH 2000.

A Century of U.S. Water Chlorination and Treatment: MMW Report; CDC 1999

Identifying Future Drinking Water Contaminants; The National Academies Press1999

The Quest for Pure Water *Baker M.N., Taras M.J.;* AWWA 1981

Water treatment: Principles and design *J.C. Crittenden et al;* John Wiley & Sons 2005

The Life of Sir William Ramsay *Morris Travers;* Arnold 1956

In Search of Penicillin David *Wilson;* Knopf 1976

Development of Deep-tank Fermentation Pfizer American Chemical Society 2008

Chemistry of Penicillin H.T. *Clarke (Ed.);* Princeton University Press 2015

Penicillin and cephalosporin production: A historical perspective *C. Muñiz, et al;* RLM 2007

Natural Gas: Fuel for the Twenty-First Century *Vaclav Smil;* John Wiley 2015

Energy Transition: The Future for Green Hydrogen; Wood Mackenzie 2019

The Chemical Engineer *Claudia Flavell-White*; IChemE 2011-2013

The Burning Question *M. Burners-Lee and Clark;* Profile Books 2013

The Carbon Collision Course: Australia's Emissions and Energy Policy Crisis *A. Perry;* APEPC 2019

On-line resources

World Population Prospects UN DESA 2022 (population.un.org)

CEFIC Facts and Figures 2023 (cefic.org)

IEA Achieving Net Zero Emissions by 2050 (energypost.eu)

Trends in Global CO_2 and Total Greenhouse Gas Emissions 2021 (www.pbl.nl/en/publications)

Net Zero: UK's contribution to stopping global warming 2019 (www.theccc.org.uk/publication)

IEA World energy outlook 2023 (www.iea.org/reports/world-energy-outlook-2023)

UK Greenhouse gas emissions 2021 DEEIS (www.gov.uk/government/statistics/provisional-uk-greenhouse-gas-emissions-national-statistics-2021)

Indicators of Global Climate Change 2022 (www.earth-system-science-data.net/about/news_and_press/indicators-of-global-climate-change-2022)

Office of Budget Responsibility (OBR) Fiscal risks report (obr.uk/frs/fiscal-risks-report-july-2021)

Energy Institute Statistical Review of World Energy 2023 (www.energyinst.org/statistical-review)

Noble Gas Market Size, Share & Trends Analysis Report (2020-2027) (www.grandviewresearch.com/industry-analysis/noble-gases-market)

Global Chemicals Market - Industry Trends and Forecast to 2031 (www.databridgemarketresearch.com/reports/global-chemicals-market)

Company and Industry Websites

Image credits

Cover: Dreamstime 311457243/Dreamstime 60908670/Dreamstime 174225966/JPHMB; Introduction: JPHMB/JPHMB/CF Industries; A: Wikimedia Commons/ Robert Oppenheim from Wikimedia Commons; B: JPHMB/ Smithsonian Institution /www.unwrappedproject.org; C:www.designingbuildings.co.uk/ Arktos from Wikipedia Commons; D: E. Foronda from Wikimedia Commons /Dyes publishing (Cope); E: Zeiss microscopy / BTL in CHM/ Dreamstime ID 6876643; F: Wikipedia commons/ Wikipedia Commons; G: Dyno Nobel Inc./ www.historia.ro; H: Royal Dutch Shell/ Dr.Semih Eser/Morris (Standard Oil Company)/Airbus SAS;I: www.unsplash.com/ D. Pickergill www. geograph.org.uk ; J: www.lungcancercoalition.org NSC/ Bibliothèque National de France;
K:www.graphene.manchester.ac.uk/www.thermofisher.com/; L: Dreamstime 314906029 BiancoBlue/ ultralifecorp.com; M: www.pxfuel.com /Dyes Publishing 2023/www.chemdiv.com; N: Mainland Minerals NZ /Archiv der Max-Planck-Gesellschaft/ Pressephotos Wikimedia Commons/ F.E Williams   Wikimedia Commons; O: N. Voitkevich via Pexels; P: Bettmann /Getty Images/ S.Torfinn /FAO/AP/Beinecke Library Yale University; Q: University of Texas Medical Branch/Science History Institute; R: Science History Institute/UC Berkeley; S: Museum London, Ontario / Wikimedia Commons /IMI Critical Engineering; T: Continental archive/ Goodyear archive; U: American Chemical Society/ Climate Doctors; V: Bryan Woolston/ Wellcome Library, London/ Piroschka van de Wouw/ANP;W: London metropolitan SWNS/Sea Cow in Wikimedia Commons; X: HELLA GmbH & Co. /Linde; Y: Alan Stones, Eli Lilly (Dista Products)/Bristol-Myers Squibb/Thomson Reuters; Z:Dreamstime 317077733 / Royal

Institution/ NASA GISS/Toyota FCV; Epilogue: Istituto Italiano di Tecnologia / www.123rf.com / JPHMB collection.

About the Author

John Blackie spent 30 years in the chemical, dye and pharmaceutical industries. Having graduated in engineering science with chemical engineering honours, he began his career in ICI at Grangemouth in central Scotland as a chemical engineer and rapidly progressed into operations management. He held senior manufacturing and business management appointments in Scotland, Manchester, West Yorkshire, France, Belgium and Teesside. He then returned to Scotland as a site manager and manufacturing director. He was a fellow of the Institute of Chemical Engineers and a member of the Board of Management of Forth Valley College.

In later years he studied horticulture and retrained in heritage gardening, working within the National Trust for Scotland and several large estate gardens and ran a gardening consultancy and maintenance business for 12 years. He lives in Perthshire with his wife who teaches chemistry at a secondary school. He has had a long-time interest in industrial history, vintage electronics and has written several articles and books on the history of the chemical industry in Scotland.